Future of Work in Asia

Chorng Yuan Fung · Agnes Siang Siew Lim ·
Weng Marc Lim · Huey Fen Cheong
Editors

Future of Work in Asia

Transforming Work Models, Workforce Dynamics and Workplace Technology

Springer

Editors
Chorng Yuan Fung
Swinburne University of Technology
Sarawak, Malaysia

Agnes Siang Siew Lim
Universiti Malaysia Sarawak
Sarawak, Malaysia

Weng Marc Lim
Sunway Business School
Sunway University
Selangor, Malaysia

Huey Fen Cheong
Universiti Malaya
Kuala Lumpur, Malaysia

ISBN 978-3-031-95803-8 ISBN 978-3-031-95804-5 (eBook)
https://doi.org/10.1007/978-3-031-95804-5

This Springer imprint is published by the registered company Springer Nature Switzerland AG
The registered company address is: Gewerbestrasse 11, 6330 Cham, Switzerland

Titles on the *Futures of Asia*

Future of Business in Asia: Exploring Pathways to Tomorrow's Business Landscape
Future of Cities in Asia: Future Direction of Urban Planning, Technology and Environmental Sustainability
Future of Economy in Asia: Geopolitical Diplomacy, Technological Revolutions and Sustainable Pathways
Future of Education in Asia: Sustainability, Technology and Internationalisation
Future of Healthcare in Asia: Advancing Equity, Ethics and Technology in Asian Healthcare
Future of Media in Asia: Artificial Intelligence, Digital Technology and Media Practice
Future of Work in Asia: Transforming Work Models, Workforce Dynamics and Workplace Technology

Acknowledgements

The editors thank all of those who contributed to the completion of this book and the *Futures of Asia* series that comprises seven volumes, including this one.

We extend our deepest appreciation to the chapter authors for their hard work and kind cooperation throughout the process. We are also grateful to the individuals and organisations that helped disseminate the calls for papers (CFPs) like the Regional Studies Association (RSA), various academic research groups on social media (particularly Facebook), and our respective institutions, including our faculties and international relations offices. Special thanks go to our assistant, Fatin Najihah, for her invaluable support in the early stages of this project.

We are also deeply grateful for the constructive feedback provided by the peer reviewers of this book, as well as the opportunities, guidance and support from Springer's commissioning editor, Doris Bleier.

Additionally, we also acknowledge and thank the following scholars for their insightful feedback on individual chapter manuscripts, both published and unpublished in this volume:

Reviewed 2 Manuscripts

1. Abdul Hayy Haziq Mohamad (University of Technology Sarawak, Malaysia)
2. Aries Yulianto (Universitas Pembangunan Jaya, Indonesia)
3. Daisy Mui Hung Kee (Universiti Sains Malaysia, Malaysia)
4. Marc Arul Weissmann (Sunway University, Malaysia)
5. Sueh Ing Su (Swinburne University of Technology Sarawak Campus, Malaysia)
6. Surbhi Pahuja (Dominion University College, Ghana)
7. Wenjing Huang (Independent researcher, USA)

Reviewed 1 Manuscript

1. Agatha Lamentan Muda (Universiti Malaysia Sarawak, Malaysia)
2. Faizan Iza Binti Zainuddin (Universiti Malaysia Sarawak, Malaysia)
3. Harsh Mahaseth (O. P. Jindal Global University, India)
4. Ida Fatimawati Adi Badiozaman (Swinburne University of Technology Sarawak Campus, Malaysia)
5. Jiankun Gong (Universiti Malaya, Malaysia)
6. Mung Ling Voon (Swinburne University of Technology Sarawak Campus, Malaysia)
7. Sora Yoo (North Carolina State University, USA)
8. Stephen T. Homer (Sunway University, Malaysia)
9. Tze-Yin Lim (Swinburne University of Technology Sarawak Campus, Malaysia)

We also thank peer reviewers who opted for anonymity.

Again, we extend our deepest gratitude for the support, time, and expertise of these individuals. Their contributions were instrumental in publishing this book. In case we have unintentionally overlooked any individual or organisation that contributed to this project, we express our sincere appreciation and gratitude as well.

Contents

List of Figures

List of Tables

About the Editors

Chorng Yuan Fung is the Head of School (Business) at the Faculty of Business, Design and Arts, Swinburne University of Technology Sarawak Campus, Malaysia. He has over 25 years of academic and industry experience as a chartered accountant. Fung is currently a member of the Association of Chartered Certified Accountants (ACCA) Malaysia Advisory Committee and the Malaysian Institute of Accountants Sarawak Regional Committee. He holds a Ph.D. in Education and an M.Sc. in human resource development. His research interests include corporate reporting, human resource development, and higher education pedagogy. Fung also supervises postgraduate students and regularly contributes to several high-impact disciplinary journals.

Agnes Siang Siew Lim is a Human Resource Development (HRD) expert at the Faculty of Cognitive Sciences and Human Development, Universiti Malaysia Sarawak (UNIMAS). She served as an Academic Coordinator and is currently the Program Coordinator for the M.Sc. HRD postgraduate program. She has experience in multinational organisations and has led training system enhancement and continuous improvement (Kaizen). Her research focuses on HRD and TQM, and she has received various professional accolades and certifications such as Accredited Trainer, Certified Mentor, and Lean Green Belt. Actively engaged in community and research projects, she contributes thought leadership through media publications and as a speaker at international seminars.

Weng Marc Lim is a Distinguished Professor and the Dean of Sunway Business School at Sunway University, an ASU-Cintana Alliance Global Partner Affiliate Faculty at Arizona State University in the United States of America, and an Adjunct Professor at Sungkyunkwan University and Swinburne University of Technology's home campus in Melbourne, Australia, and international branch campus in Sarawak, Malaysia, where he served as Dean and Head of School as well as on the University Council. Professor Lim is the Record Holder for "Youngest Professor" in the Malaysia Book of Records and features in the Academy of Sciences Malaysia's Top Research Scientists Malaysia and JCI's Ten Outstanding Young Malaysians. With over 100 publications in prestigious “A*” and “A” ranked journals, Professor Lim’s research has been cited more than 40,000 times. As a thought leader, Professor Lim has led high-level discussions at AppliedHE, Association to Advance Collegiate Schools of Business (AACSB), Association of MBAs (AMBA), Business Graduates Association (BGA), Quacquarelli Symonds (QS), Times Higher Education (THE), United Nations Educational, Scientific and Cultural Organization (UNESCO) and Principles of Responsible Management Education (UNPRME), Nikkei Forum, and World Economic Forum (WEF). Professor Lim holds a Bachelor of Business and Commerce with First Class Honors and a Doctor of Philosophy from Monash University, and multiple certificates post-Ph.D., including a Graduate Certificate in Learning and Teaching (Higher Education) with High Distinction from Swinburne University of Technology and a Certificate in School Management and Leadership from Harvard University.

Huey Fen Cheong is a Senior Lecturer at Universiti Malaya, Malaysia. Her interests have been multidisciplinary, mainly linguistics, sociology, education, marketing, and psychology. Driven by a cultural and postcolonial perspective, she collaborated with a group of established scholars to initiate this *Futures of Asia* series (which contains this book) and the earlier *COVID-19 in Malaysia* series. These initiatives aim to promote Global South research and publications, especially on significant events such as the COVID-19 pandemic, post-pandemic normality and transformations, and the rise of AI, which equally matter to non-Western regions and deserve equal scholarly attention (and acceptance) beyond Western-centric frameworks in academia.

Editors' Note

The clusters of future studies in each chapter are based on Fergnani (2019, 2020), who identified six (6) key research/thematic clusters from reviews of future studies since 1968: "corporate foresight; past and futures; humanity at the limen; environmental futures; post-normality and complexity, and technological trends". While the chapter topics present different aspects of work in Asia, the clusters of future studies show different dimensions of futures and future studies.

References

Fergnani, A. (2019). Mapping futures studies scholarship from 1968 to present: A bibliometric review of thematic clusters, research trends, and research gaps. *Futures, 105*: 104–23.

Fergnani, A. (2020). Mapping fifty years of futures studies scholarship (1968–2017). In R. Slaughter & A. Hines (Eds.), *The Knowledge Base of Futures Studies 2020* (pp. 24–47). US: Association of Professional Futurists and Foresight International.

Chapter 1
Introduction: The Future of Work Models, Work Dynamics and Workplace Technology in Asia

Chorng Yuan Fung and Agnes Siang Siew Lim

Abstract The future of work is one of the most pressing topics of the twenty-first century, with technological advancements, shifting social dynamics, and changing economic landscapes driving fundamental transformations in workplaces across the globe. However, the trajectory of work in Asia is distinct, shaped by unique cultural, economic, and demographic factors that set the region apart from the rest of the world. The rapid advancement of information technology, increasing awareness of mental well-being, environmental, social and governance (ESG) principles, and femininity of work culture all have impacted the work models, work dynamics, and the work environment. Various work models emerged, causing workplace dynamics to be more accommodative, and we saw an influx of new workplace technology being deployed to support the changes in the work models. This chapter aims to provide an overview of this book, covering these aspects of the work landscape based on literature and attempts to project the future of work in Asia. This knowledge is essential to equip policymakers, business leaders, and employees with essential perspectives on adapting to evolving dynamics and fostering a more inclusive, resilient, and sustainable future of work in Asia.

Keywords Work models · Workplace dynamics · Workplace technology · Work landscape

---

C. Y. Fung (✉)
Swinburne University of Technology Sarawak Campus, Kuching, Malaysia
e-mail: cfung@swinburne.edu.my

A. S. S. Lim
Universiti Malaysia Sarawak, Kota Samarahan, Malaysia
e-mail: lssagnes@unimas.my

C. Y. Fung et al. (eds.), *Future of Work in Asia*,
https://doi.org/10.1007/978-3-031-95804-5_1

1.1 Introduction: About the Book

This book seeks to critically examine the future of work in Asia, written by authors who bring a unique, regional perspective to these pressing issues. Pairing 'future' and 'Asia' aligns with the rising of Asia and the prevailing sentiment—"the future is Asia". This trend is also addressed by the 25th ASAA Biennial Conference 2024 (ASAA: Asian Studies Association of Australia), which explores "Asia Futures: Studies of, in and with Asia". The escalating prominence of Asia on the global stage has spurred the emergence of various sub-disciplines dedicated to the region, as seen in courses like 'Asian Economics at New York University and 'Business in China in Oxford's MBA programme. This indicates a growing academic interest in Asian-centric studies. This trajectory underscores the book's relevance and potential appeal to institutions and individuals seeking to incorporate Asian elements and/or perspectives into their curricula and research. Furthermore, certain larger Asian countries tend to receive more recognition and opportunities in general 'global' publications compared to their smaller counterparts. This book, with a specific focus on Asia, will promote the expansion of coverage across diverse Asian regions to be more representative of the continent.

Drawing on the lived experiences of scholars and researchers across Asia offers insights rooted in local realities, cultural nuances, and the socio-political contexts shaping the work environment. The region's workforce is not a monolith but a complex mosaic of diverse workforces, industries, and national policies, including perspectives from Asian authors is essential. These authors are well-positioned to explore technological advancements, work models, and new social expectations—particularly concerning well-being and the role of mothers in the workplace—to transform work dynamics in Asia.

This book critically explores the future trajectories of Asian work through a collection of empirical, conceptual, and review papers. Rooted in the understanding that future studies do not predict the real future but rather analyse ideas about it, the book follows Dator's (2019) framework, emphasising the forecasting of "alternative futures" and the continuous development of "preferred futures". The book aligns with six identified research clusters in future studies by Fergnani (2020): "corporate foresight", "past and futures", "humanity at the limen", "environmental futures", "post-normality and complexity", and "technological trends". Each chapter will be marked by the relevant research cluster(s). By examining these dimensions within the context of the Asian future work, the book aims to contribute valuable insights to the dynamic discourse on future studies while offering a nuanced understanding of the work possibilities that await Asia and deciphering Asia's evolving impact and roles within the broader global context.

As we look toward the future, this book aims to equip policymakers, business leaders, and workers with the knowledge needed to navigate a rapidly changing landscape and to build a more inclusive, resilient, and sustainable future of work in Asia.

1.2 The Meaning and Perspectives of Work

To understand 'work', we must first understand the meaning and perspectives of work. Rosso et al. (2010) did a literature review on the meaning of work. They synthesised that the meaning of work should encompass value, motivation, and belief. Workers reap the value of work that gives them desire and feeling. That value could be a product of culture, institution, or personal influence, which is context-specific. The value could be attached to the meaning of work and correlates to the motivation for work. As summarised by Rosso et al. (2010), the motivation for work is the internal feeling the workers experience when performing a task, which extrinsic and/or extrinsic factors could drive. Lastly, workers' belief in their work could determine the meaning of work. The belief could be determined by their job involvement, work centrality, work orientation, and callings. Further, Rosso et al. (2010) also synthesised that co-workers could influence the meaning of work, the leaders of the organisations, the community the workers live in, and their families. All these syntheses suggest two essential elements of the meaning of work. Firstly, it stems from the self, i.e., the intrinsic perspective. Secondly, it could stem from the extrinsic perspective, i.e., the environment in which one operates.

With this understanding in mind, we could dive into the perspective of work. There are multiple lenses for this purpose (refer Yeoman et al., 2019). In this book, we chose to look at work from a humanistic perspective. The humanistic perspective on work emphasises the workers' well-being and the impact of work within the workplace (Dierksmeier, 2016; Melé, 2016). It views individuals beyond the contributors to productivity but on their holistic and meaningful engagement and interpersonal relationships with the environment. This approach converged with our understanding of work; together, we could understand work meaningfully.

In Asia, the diverse cultural and economic transformations are rapidly changing the nature of work and its environment. The humanistic lens will provide an inclusive framework to navigate these shifts. It highlights the importance of balancing technological advancements and economic growth with the ethical imperative to uphold workers' rights, foster equitable opportunities, and ensure psychological and social well-being (Steinberg et al., 2024). Such an approach highlights the value of collaboration, adaptability, and empathy in building a resilient workforce (Mohan, 2023). This approach aligns with Asia's rich traditions of community-centric philosophies, offering pathways to create work environments where human flourishing and organisational success coexist harmoniously. By reimagining work through this lens, the region can shape a future of work that honours both economic and human imperatives.

The COVID-19 pandemic has forcefully shifted how people work, which, in some ways, also changed work dynamics (Bouzaghrane et al., 2024; Gintova, 2024). The impact was most felt in the Asian region where the outbreak started. As this region recovers from the profound disruptions caused by the COVID-19 pandemic, it has become increasingly clear that the traditional work paradigms are no longer sufficient to address the diverse challenges workers and employers face. The pandemic accelerated many pre-existing trends, such as remote work, digital transformation, and the reimagining of work-life balance (Mathur et al., 2024). Still, it also exposed deep vulnerabilities in work dynamics, especially for marginalised groups like working mothers (Aura & Desiana, 2023). The future of work in Asia will require innovative models adaptable to technological advancements and responsive to the evolving needs of a diverse workforce, including those juggling family responsibilities. Moreover, as workplace technology continues to reshape daily operations, organisations must consider the impact of automation, artificial intelligence, and digital tools on employee engagement, job security, and overall well-being (Aura & Desiana, 2023; Bhargava et al., 2021).

1.3 Book Structure and Overview

1.3.1 Part I: Future Work Models in Asia

Work models could refer to work arrangements (Hopkins & Bardoel, 2023). These models might include hybrid working, work-from-home, remote working, etc. Most work models would embed flexibility, giving the workers more control over how work is arranged (Babapour Chafi et al., 2021). However, the work models vary from one industry to another. Some industries could accommodate more flexibility than others, while other functions of jobs would have less space for flexibility (Al-Alwan et al., 2022; Babapour Chafi et al., 2021; Hopkins & Bardoel, 2023; Howard, 2017). Due to cultural factors, especially in Asia, women could face limited opportunities for a flexible working model in their employment (Abdel-Rahman et al., 2023; Shabir & Gani, 2020; Yen et al., 2024).

Among the various industries that faced the shift of the flexible working model movement, the impact was significant for the digital platform-based industry. This was due to the emergence of the gig economy since the pandemic (Shahzadi et al., 2022). Dunn et al. (2023) investigated multiple types of platform workers concerning their perception of flexibility. They found that different platform types and how they mediate flexibility led to important differences in how workers experience their work. However, articles that provide insight into the work models of platform workers, especially in the Asian region, are scarce (Au-Yeung et al., 2024; Dunn et al., 2023; Wu et al., 2022). As the Asian region operates in a diverse cultural, social, technological, and political environment, it is essential to have a consolidated view of how the work on the platforms is operated, showing the differences in these various aspects. Such insight would be valuable to policymakers and employers in formulating inclusive policies and procedures to embrace workforce diversity. In Chap. 2, Nazruzila Razniza Mohd Nadzria and Nusrah Mamatb from Open University Malaysia, and Sharmila Devi from INTI International University, Nilai, Malaysia, attempt to provide insights into platform workers, explore a representation framework of some Asian countries, offering a diverse perspective on the framework for platform workers for the future Asian work landscape.

The contemporary movement of Environmental, Social and Governance (ESG) also shapes the future of work in Asia. The landscape for ESG has emerged over the past decades, with more awareness and initiatives embedding the ESG principles (Pillai et al., 2024). Work models are no exception. Stakeholders perceive organisations that embrace ESG as good corporate citizens, often making them the preferred employers for graduates (Hoppe et al., 2022; Khamisu et al., 2024; Ng et al., 2020). Hence, there is a need to incorporate ESG into the human resource strategy. ESG needs to be embedded as part of the corporate culture to be effective. One of the challenges is the measurement of ESG in the work models. Past literature offers limited guidance on developing a measurement matrix that could help organisations assess the impact of ESG on the work models in the Asian context. In Chap. 3, Jing Yi Yong, Woon Leong Lin, Thanuja Rathakrishnan, Thivashini B. Jaya Kumar, Feranita Feranita and Nelvin XeChung Leow, from Taylor's University, Malaysia, provide insights into the incorporation of ESG into the human resource strategy in the Asian context. They propose measuring matrices of ESG on human resource functions that could be helpful for corporates and researchers in this area in the future Asian work landscape.

1.3.2 Part II: Future Work Dynamics in Asia

When we dive into the Asian work environment, we see a diverse culture of races, languages, and backgrounds (Zhang, 2023). These factors influence the work dynamics, and it is challenging to examine them in isolation (Campton et al., 2023). All these variables interact with each other reciprocally. Over the past five years, the work dynamics have become more complex with Generation Z's entry into the work landscape. Their existence poses challenges and tension to the landscape managed by Generation X, and Y. Gen Z has significant differences in lifestyle and culture stemming from their growing up in the technological era and improving quality of life (Tanveer et al., 2020). To ensure harmonious workplace dynamics, these generations must have a sound understanding of each other. Past literature that consolidates the characteristics, working preferences, communication styles, preferred leadership styles, and rewards are few. Most studies examine these variables in isolation. Hence, it is crucial to have an article that provides these insights. In Chap. 4, Sheena D.C. Doria-De Vera from the University of Philippines, Philippines, using typology as a research methodology, provides an overview of the work landscape of a multi-generational workforce. The results are essential to help policymakers and business leaders develop an inclusive culture accommodating the expectations of the multigenerational workforce in the future Asian work environment.

Since the COVID-19 pandemic, mental well-being has been a constant challenge at the workplace. Stress that comes from not only work but also the pressure expected and relationships with colleagues could trigger mental health issues (Wu et al., 2023). In the Asian context, where the culture is more conservative and less vocal, collectively, it would pose challenges for better mental health diagnosis and management

(Au, 2023; Huang et al., 2022). However, past literature that reports some of these issues more inclusively is rare (de Oliveira et al., 2023). Most studies examined these factors separately (Ran et al., 2021). Hence, it would be helpful to compress all these variables to give the readers a broad yet holistic view and to comprehend their inter-relationships, especially in the Asian context. In Chap. 5, Kiat Sing Heng, Chia Hua Sim and Mei Yuan Law, from Swinburne University of Technology, Sarawak Campus, Malaysia, paint the pathway from traditional work values to contemporary post-COVID-19 work values, highlighting the increasing mental health issues in the workforce. It is evident that culture and mental well-being are intertwined and must be managed holistically. They provide some suggestions for business leaders to ensure the mental well-being of the workforce is intact, charting a way forward to a (mental) healthier future Asian workforce.

Women are another key player in the workplace. Their role cannot be undermined. After the COVID-19 pandemic, many women could enter the workplace due to the emergence of flexible work arrangements (FWA). However, Asian countries are moving toward FWA less progressively. In many organisations, women are marginalised (Setyonaluri & Utomo, 2024). Policies do not sufficiently protect women's rights, rewards, and equality (Prasetyo, 2024). In Chap. 6, Donna Barclay, Nivakan Sritharan and Jacob Ting King Soon from Swinburne University of Technology Sarawak Campus, Malaysia, report on the Asian work landscape for working mothers from the perspective of employment inclusivity and human resource policies. Their findings fill the gaps in the literature, providing thought-provoking knowledge for policymakers and business leaders to reconsider the current policies and employment models, which are crucial for the future Asian work landscape.

1.3.3 Part III: Future Workplace Technology in Asia

The evolution of technology has been rapid in the past decade, transforming social and working lives. Virtually, most of the work would involve technology to some extent. Digital literacy has become a basic skill that all workers need to equip themselves with (Morris & Rohs, 2021; Reddy et al., 2022). Past studies have examined digital literacy with employability and career success. However, every industry has unique characteristics that offer different landscapes for career advancement (Fung & Gunasekara, 2021). There is a need to look at this topic across different industries to provide a holistic view of its impact. Furthermore, such an attempt could enable policymakers at the national level to formulate strategies and policies that could elevate the quality of the national workforce and further propel the nation's competitiveness in attracting foreign investments. In Chap. 7, Chow Poh Ling, Phuah Kit Teng, Bernard Lim Jit Heng, Ow Mun Waei and Khoong Tai Wai from Tunku Abdul Rahman University of Management and Technology, Malaysia, and Siti Intan Nurdiana Wong Abdullah from Nottingham Trent University, United Kingdom, reported that in some Asian countries, workers with high proficiency in digital tools and technologies tend to have better job security, higher career advancement potential, and are more adaptable to industry changes. This chapter explores these empirical findings in some Asian countries, bridges the research gap, and provides insights into training for upskilling and other developmental initiatives in the future Asian work landscape.

Technology offers an approach that saves time and resources, especially in a professional setting (Harris et al., 2020). Architects use 3D modelling, lawyers have access to AI for legal solutions, and auditors have Computer-Assisted Audit Techniques (CAAT) to conduct the audit more efficiently and effectively. However, with the evolution of Industry Revolution 4.0 (IR 4.0), these technologies need to be reviewed and revised to be integrated with the IR 4.0 systems (Candra et al., 2021). In Chap. 8, Dayana Jalaludin, Suhaily Shahimi and Ravinder Kaur A/P Harminder Singh from Malaya University, Malaysia, analyse past literature and secondary data, and by adopting the Technology, Organization, and Environment (TOE) framework (Tornatzky et al., 1990), propose the building blocks for audit efficiency with advanced Computer Assisted Audit Tools and Techniques (CAATTs) adoption. Such an approach integrated the framework tailored to the fourth industrial revolution (IR4.0), encompassing technological, organisational, and environmental considerations crucial for the future Asian work landscape, particularly in the auditing profession.

COVID-19 has pushed society to a New Way of Working (NWW). The extensive use of technology has transformed the work models and dynamics to a new frontier. However, technology has been intrusive and disruptive, especially in Asia, where technology is seen as emerging (Chong et al., 2024; Schmidthuber et al., 2020). Society has slowly recognised technology as a source of social illness (Harahap et al., 2023). The existence of potential relationships between the NWW, the use of technology, and personal and organisational level factors suggested that organisations need suitable initiatives to strike a balance between the use of technology and workers' productivity (Yener et al., 2021). In Chap. 9, Nguyen Ngoc Duy Phuong from the International University and Vietnam National University, Vietnam, Duong Cong Hiep and Giang Thi Trang, both from Chang Jung Christian University, Taiwan and Thai Nguyen University of Economics and Business Administration, Vietnam, and Nguyen Hong Huan from Maharishi International University, Iowa, USA, explores the relationship between employee well-being, turnover intention, and technostress in some Asian countries. Their findings highlight the complex relationship between these variables and the role of culture in explaining the relationship. This study further emphasises the importance of suitable interventions to mitigate the technostress arising from the NWW, which is the trend of the future work landscape.

1.4 The Way Forward

The future work landscape in Asia is dynamic and subject to many influences, mainly from emerging technologies, national and organisational culture, and geopolitical influence (Lim et al., 2022). Organisations and business leaders must ensure the work ecosystem is responsive to societal and technological changes. At the same time, the workers must have a full knowledge of environmental changes and take control of their career development for success (Fung & Gunasekara, 2021). The future workers in Asia must be agile and vigilant as they enter the dynamic future work landscape. This landscape offers rich potential for research and space for innovative endeavours for a better future.

To provide a trajectory perspective of the future of work in Asia, the authors of each chapter have carefully articulated a section on 'Future Direction' and 'Highlights' towards the end of each chapter. These sections are essential not only to consolidate but also to summarise the topic discussed. We hope readers will find some of these points relevant and put them into action.

References

Abdel-Rahman, S., Abonazel, M. R., Awwad, F. A., & Kibria, B. G. (2023). The impact of COVID-19-induced responsibilities on women's employment in Arab countries. *Sustainability, 15*(13), 9856. https://doi.org/10.3390/su15139856

Al-Alwan, M., Bader, D., Al-Qatawneh, M., Alneimat, S., & Al-Hawary, S. (2022). E-HRM and employee flexibility in Islamic banks in Jordan. *International Journal of Data and Network Science, 6*(3), 703–710. https://doi.org/10.5267/j.ijdns.2022.4.002

Asian Studies Association of Australia. (2024). *25th Biennial Asian studies association of Australia conference: Asia futures: Studies of, in and with Asia* (2024, July 1–4). Curtin University, Perth, Western Australia. https://www.asaa2024.org/home

Au, A. (2023). *Mental health in East Asia: Cultural beliefs, social networks, and mental health experiences*. Routledge. https://doi.org/10.4324/9781003308720

Aura, N. A. M., & Desiana, P. M. (2023). Flexible working arrangements and work-family culture effects on job satisfaction: The mediation role of work-family conflicts among female employees. Jurnal Manajemen Teori Dan Terapan. *Journal of Theory and Applied Management, 16*(2). https://doi.org/10.20473/jmtt.v16i2.45960

Au-Yeung, T. C., Chan, C. K. C., Ming, C. K. K., & Tsui, W. Y. A. (2024). The gig economy, platform work, and social policy: food delivery workers' occupational welfare dilemma in Hong Kong. *Journal of Social Policy*, 1–19. https://doi.org/10.1017/S0047279423000673

Babapour Chafi, M., Hultberg, A., & Bozic Yams, N. (2021). Post-pandemic office work: Perceived challenges and opportunities for a sustainable work environment. *Sustainability, 14*(1), 294. https://doi.org/10.3390/su14010294

Bhargava, A., Bester, M., & Bolton, L. (2021). Employees' perceptions of the implementation of robotics, artificial intelligence, and automation (RAIA) on job satisfaction, job security, and employability. *Journal of Technology in Behavioral Science, 6*(1). https://doi.org/10.1007/s41347-020-00153-8

Bouzaghrane, M. A., Obeid, H., González, M., & Walker, J. (2024). Human mobility reshaped? Deciphering the impacts of the Covid-19 pandemic on activity patterns, spatial habits, and schedule habits. *EPJ Data Science, 13*(1). https://doi.org/10.1140/epjds/s13688-024-00463-4

Campton, J., Tham, A., & Ting, H. (2023). Quiet quitting–implications for Asian businesses. *Asian Journal of Business Research Volume, 13*(2), 128–134. https://doi.org/10.14707/ajbr.230153

Candra, R., Wahyuni, E., & Shobirin, K. (2021). Accountant profession: Function, role and his order during the industrial revolution 4.0. *Jurnal Ilmiah Manajemen, Ekonomi, & Akuntansi (MEA), 5*(1), 534–551. https://doi.org/10.31955/mea.v5i1.698

Chong, A. S. E., Fung, C. Y., & Ki, Y. P. (2024). The relationship of e-training, work motivation, and work-life balance on Malaysian teachers' performance during COVID-19 pandemic. *Malaysian Journal of Business and Economics, 11*(1). https://doi.org/10.51200/mjbe.v11i1.5292

Dator, J. (2019). What futures studies is, and is not. *Jim Dator: A noticer in time: Selected work, 1967–2018* (pp. 3–5). Springer International Publishing.

de Oliveira, C., Saka, M., Bone, L., & Jacobs, R. (2023). The role of mental health on workplace productivity: A critical review of the literature. *Applied Health Economics and Health Policy, 21*(2), 167–193. https://doi.org/10.1007/s40258-022-00761-w

Dierksmeier, C. (2016). What is 'humanistic' about humanistic management? *Humanistic Management Journal, 1*, 9–32. https://doi.org/10.1007/s41463-016-0002-6

Dunn, M., Munoz, I., & Jarrahi, M. H. (2023). Dynamics of flexible work and digital platforms: Task and spatial flexibility in the platform economy. *Digital Business, 3*(1), Article 100052. https://doi.org/10.1016/j.digbus.2022.100052

Fergnani, A. (2020). Mapping fifty years of futures studies scholarship (1968–2017). In R. A. Slaughter, A. Hines (Eds.), *The knowledge base of futures studies 2020* (pp. 24–47). US: Association of Professional Futurists and Foresight International

Fung, C. Y., & Gunasekara, A. (2021). Re-thinking post COVID-19 career success: Insights from contemporary career approaches. In W. M. Lim, S. Kaur, & H. F. Cheng (Eds.), *COVID-19, business, and economy in Malaysia: Retrospective and prospective perspectives*. Routledge, UK. https://doi.org/10.4324/9781003182740-9

Gintova, M. (2024). Public Service in Canada post-COVID-19 pandemic: Transitioning to hybrid work and its implementation challenges. *Canadian Public Administration, 67*(1). https://doi.org/10.1111/capa.12550

Harahap, M. A. K., Sutrisno, S., Fauzi, F., Jusman, I. A., & Ausat, A. M. A. (2023). The impact of digital technology on employee job stress: A business psychology review. *Jurnal Pendidikan Tambusai, 7*(1), 3635–3638. https://doi.org/10.31004/jptam.v7i1.5775

Harris, M., Riley, J., & Venkatesh, R. (2020). Psychological capital and robotic process automation: Good, bad, or somewhere in-between? *Journal of Emerging Technologies in Accounting, 17*(1), 71–76. https://doi.org/10.2308/jeta-52650

Hopkins, J., & Bardoel, A. (2023). The future is hybrid: How organisations are designing and supporting sustainable hybrid work models in post-pandemic Australia. *Sustainability, 15*(4), 3086. https://doi.org/10.3390/su15043086

Hoppe, D., Keller, H., & Horstmann, F. (2022). Got employer image? How applicants choose their employer. *Corporate Reputation Review*, 1–21. https://doi.org/10.1057/s41299-021-00119-3

Howard, J. (2017). Nonstandard work arrangements and worker health and safety. *American Journal of Industrial Medicine, 60*(1), 1–10. https://doi.org/10.1002/ajim.22669

Huang, Z., Shi, Y., & Wang, Y. (2022). Does growth mindset benefit mental health in Asia? Evidence from Chinese students. *Journal of Pacific Rim Psychology, 16*, 18344909221135360. https://doi.org/10.1177/18344909221135358

Khamisu, M. S., Paluri, R. A., & Sonwaney, V. (2024). Stakeholders' perspectives on critical success factors for environmental social and governance (ESG) implementation. *Journal of Environmental Management, 365*, Article 121583. https://doi.org/10.1016/j.jenvman.2024.121583

Lim, W. M., Chin, M. W. C., Ee, Y. S., Fung, C. Y., Giang, C. S., Heng, K. S., Kong, M. L. F., Lim, A. S. S., Lim, B. C. Y., Lim, R. T. H., Lim, T. Y., Ling, C. C., Mandrinos, S., Nwobodo, S., Phang, C. Su. C., She, L., Sim, C. H., Su, S. I., Wee, G. W. En., & Weissmann, M. A. (2022). What is at stake in a war? A prospective evaluation of the Ukraine and Russia conflict for business and society. *Global Business and Organizational Excellence, 41*(6), 23–36. https://doi.org/10.1002/joe.22162

Mathur, G., Gulati, C., Sankpal, S., & Kee, D. M. H. (2024). Work-life balance and withdrawal behaviors among knowledge workforce: High performance work system comes into play for managing boundaries. *Global Business and Organizational Excellence, 43*(6), 24–40. https://doi.org/10.1002/joe.22257

Melé, D. (2016). Understanding humanistic management. *Humanistic Management Journal, 1*, 33–55. https://doi.org/10.1007/s41463-016-0011-5

Mohan, K. P. (2023). Mental health and well-being support for Thai university graduates: A qualitative exploration of pathways to develop a resilient workforce. *Journal of Population and Social Studies, 31*. https://doi.org/10.25133/JPSSV312023.043

Morris, T. H., & Rohs, M. (2021). Digitization bolstering self-directed learning for information literate adults—A systematic review. *Computers and Education Open, 2*, Article 100048. https://doi.org/10.1016/j.caeo.2021.100048

Ng, T. H., Lye, C. T., Chan, K. H., Lim, Y. Z., & Lim, Y. S. (2020). Sustainability in Asia: The roles of financial development in environmental, social and governance (ESG) performance. *Social Indicators Research, 150*, 17–44. https://doi.org/10.1007/s11205-020-02288-w

Pillai, R., Islam, M. A., Sreejith, S., & Al-Malkawi, H. A. (2024). Comparative analysis of environmental, social and governance (ESG) ratings: do sectors and regions differ? *Journal of Management and Governance*, 1–41. https://doi.org/10.1007/s10997-023-09692-7

Prasetyo, D. A. (2024). Legal protection for women workers in Indonesia: A critical review of law application. *Pena Justisia: Media Komunikasi dan Kajian Hukum, 23*(3), 48–60. https://doi.org/10.31941/pj.v23i3.5233

Ran, M. S., Hall, B. J., Su, T. T., Prawira, B., Breth-Petersen, M., Li, X. H., & Zhang, T. M. (2021). Stigma of mental illness and cultural factors in Pacific Rim region: A systematic review. *BMC Psychiatry, 21*, 1–16. https://doi.org/10.1186/s12888-020-02991-5

Reddy, P., Sharma, B., & Chaudhary, K. (2022). Digital literacy: A review in the South Pacific. *Journal of Computing in Higher Education, 34*(1), 83–108. https://doi.org/10.1007/s12528-021-09280-4

Rosso, B. D., Dekas, K. H., & Wrzesniewski, A. (2010). On the meaning of work: A theoretical integration and review. *Research in Organizational Behavior, 30*, 91–127. https://doi.org/10.1016/j.riob.2010.09.001

Schmidthuber, L., Maresch, D., & Ginner, M. (2020). Disruptive technologies and abundance in the service sector-toward a refined technology acceptance model. *Technological Forecasting and Social Change, 155*, Article 119328. https://doi.org/10.1016/j.techfore.2018.06.017

Steinberg, M., Zhang, L., & Mukherjee, R. (2024). Platform capitalisms and platform cultures. *International Journal of Cultural Studies*. https://doi.org/10.1177/13678779231223544

Setyonaluri, D., & Utomo, A. (2024). Negotiating work, family, and traffic: Articulations of married women's employment decisions in Greater Jakarta. *Gender, Work & Organization, 31*(6), 2423–2445. https://doi.org/10.1111/gwao.13069

Shahzadi, I., Rafiq, S., & Ali, U. (2022). Investigating the influence of flexible work arrangements on work-life balance in South Asian gig workers: Does Ryff's six-factor model of psychological well-being moderates? iRASD *Journal of Management, 4*(2), 316–329. https://doi.org/10.52131/jom.2022.0402.0081

Shabir, S., & Gani, A. (2020). Impact of work–life balance on organizational commitment of women health-care workers: Structural modeling approach. *International Journal of Organizational Analysis, 28*(4), 917–939. https://doi.org/10.1108/IJOA-07-2019-1820

Tanveer, M., Hassan, M., Shah, Z. A., & Islam, S. U. (2020). Clash of generations: Assessing the impact of generational diversity in a workforce. *International Journal of Business and Economic Affairs, 5*(4), 184–198. https://ijbea.com/ojs/index.php/ijbea/article/view/185

Tornatzky, L. G., Fleischer, M., & Chakrabarti, A. K. (1990). *The processes of technological innovation*. Lexington Books.

Wu, P. F., Zheng, R., Zhao, Y., & Li, Y. (2022). Happy riders are all alike? Ambivalent subjective experience and mental well-being of food-delivery platform workers in China. *New Technology, Work and Employment, 37*(3), 425–444. https://doi.org/10.1111/ntwe.12243

Wu, Q., Cao, H., & Du, H. (2023). Work stress, work-related rumination, and depressive symptoms in university teachers: Buffering effect of self-compassion. *Psychology Research and Behavior Management, 16*, 1557–1569. https://doi.org/10.2147/PRBM.S403744

Yen, W. T. M., Yen, Y. Y., Ming-Li, T., & Ying, T. W. (2024). Empowering women in the workplace: Assessing female employees' readiness for flexible work arrangements. *Asian Economic and Financial Review, 14*(2), 143. https://doi.org/10.55493/5002.v14i2.5011

Yener, S., Arslan, A., & Kilinç, S. (2021). The moderating roles of technological self-efficacy and time management in the technostress and employee performance relationship through burnout. *Information Technology & People, 34*(7), 1890–1919. https://doi.org/10.1108/ITP-09-2019-0462

Yeoman, R., Bailey, C., Madden, A., & Thompson, M. (Eds.). (2019). *The Oxford handbook of meaningful work.* Oxford University Press.

Zhang, L. (2023). Racial inequality in work environments. *American Sociological Review, 88*(2), 252–283. https://doi.org/10.1177/00031224231157303

Chorng Yuan Fung is the Head of School (Business) at the Faculty of Business, Design and Arts, Swinburne University of Technology Sarawak Campus, Malaysia. He has over 25 years of academic and industry experience as a chartered accountant. Fung is currently a member of the Association of Chartered Certified Accountants (ACCA) Malaysia Advisory Committee and the Malaysian Institute of Accountants Sarawak Regional Committee. He holds a Ph.D. in Education and an M.Sc. in human resource development. His research interests include corporate reporting, human resource development, and higher education pedagogy. Fung also supervises postgraduate students and regularly contributes to several high-impact disciplinary journals.

Agnes Siang Siew Lim is a Human Resource Development (HRD) expert at the Faculty of Cognitive Sciences and Human Development, Universiti Malaysia Sarawak (UNIMAS). She served as an Academic Coordinator and is currently the Program Coordinator for the MSc HRD postgraduate program. She has experience in multinational organisations and has led training system enhancement and continuous improvement (Kaizen). Her research focuses on HRD and TQM, and she has received various professional accolades and certifications such as Accredited Trainer, Certified Mentor, and Lean Green Belt. Actively engaged in community and research projects, she contributes thought leadership through media publications and as a speaker at international seminars.

Part I
Future Work Models in Asia

Chapter 2
Exploring Future Representation Framework for Platform Workers

Clusters of Future Studies: Past and Futures; Humanity at the Limen

Nazruzila Razniza Mohd Nadzri, Nusrah Samat, and Sharmila Devi

Abstract The rise of platform employment in Asia's dynamic labour market and economy has sparked discussions on how to represent platform workers, who often face precarious working conditions despite the opportunities the platform work presents. Reinventing the traditional concept of collective representation to better fit the characteristics of platform workers may serve as an effective means of mitigating their precarious working conditions. This chapter investigates the challenges the platform workers encounter in their collective representation and evaluates various representation frameworks across Asian countries, particularly Malaysia, Singapore, South Korea, and Japan. By comparing these Asian frameworks among themselves and with those in Western countries, the research aims to provide diverse insights into effective representation frameworks for platform workers. Using a doctrinal legal study approach, the research analyses both legal and non-legal documents, focusing on empirical practices within the existing Asian countries legal frameworks. The findings suggest that an inclusive and flexible representation approach, bolstered by legislative support, is essential to address the unique characteristics of platform work. Ultimately, this research contributes to the understanding of collective representation frameworks tailored to the specific needs of platform workers, drawing on best practices from various Asian contexts.

Keywords Platform workers · Worker representation · Collective representation · Representation framework · Precarious condition

N. R. M. Nadzri (✉) · N. Samat
Faculty of Business and Management, Open University Malaysia, Petaling Jaya, Malaysia
e-mail: nazruzila@oum.edu.my

N. Samat
e-mail: nusrah@oum.edu.my

S. Devi
Faculty of Business and Communications, INTI International University, Nilai, Malaysia
e-mail: sharmila.devi@newinti.edu.my

C. Y. Fung et al. (eds.), *Future of Work in Asia*,
https://doi.org/10.1007/978-3-031-95804-5_2

2.1 Introduction

Digital platforms have gained growing importance in facilitating transactions between buyers and sellers of products and services. Despite being initially introduced in Western markets, companies like Uber and Grab quickly redirected their attention towards acquiring workforces from emerging economies (Galpaya et al., 2023). Galpaya et al. (2023) further observed that the change was motivated by the reduced employment expenses in these nations and the increasing accessibility of internet connectivity. The majority of emerging Asian economies currently host both global and local platforms that facilitate a wide range of services, including the assignment of professional labour, ride-hailing services, and the acquisition of odd jobs (Galpaya et al., 2023). This chapter, however, specifically concentrates on two categories of platform work, which implies the context of the platform workers referred to: services provided by rideshare platforms and food delivery platforms. These platforms are meant for low-skilled activities that require the worker to be present and are described as location-based-on-demand work (De Stefano, 2015; Rafelis de Broves et al., 2024).

In contrast to web-based platforms that involve freelance platforms and micro-task platforms, location-based-on-demand platforms are commonly associated with hazardous working circumstances. These platforms often need lower qualifications, pay workers by the hour, and promote less direct communication between the worker and the customer (Cruz & Gameiro, 2023). Several platforms have implemented a contractual model that compels workers to function as independent contractors, devoid of any ability to negotiate (Forsyth, 2023). Hence, platform workers are exposed to various risks, such as receiving inadequate payment, facing significant safety hazards, enduring customer abuse, and being exploited in other ways. Additionally, the platform has the power to arbitrarily terminate workers' services through a deactivation system. Scholars have conducted a substantial amount of research to address this issue (Doherty & Franca, 2020), primarily focusing on changing the employment classification of platform workers through legal means or by reconsidering the current legislative structures (Durri, 2021). However, the collective dimension of platform work has been subjected to less scrutiny (Doherty & Franca, 2020; Forsyth, 2023). For example, the collective strategy's potential to improve the working conditions of platform workers.

Nevertheless, Asian countries like Japan, South Korea, Singapore and Malaysia have shown an increasing recognition of the importance of implementing this strategy, despite encountering different trends and practices (Araki, 2021; Trenerry et al., 2022; Uchiyama et al., 2022; Umer, 2021; Yang & Joo Chae, 2020). However, it is also observed that Malaysia for example still lags behind in exploring the potential of representative techniques and bargaining power to address the precarious conditions experienced by platform workers (Md Radzi et al., 2022). In the Western context, the collective representation among platform workers revealed a centralisation in the demand for stricter regulation over the platform model, as their regulatory systems are the most developed (Umney et al., 2024). This chapter examines instances of effective regulatory regimes and representation techniques used by platform workers in different Asian and Western countries to fill the knowledge gap.

The chapter employs doctrinal legal research, which consists of a descriptive and detailed analysis of legal sources, including legislation and case law. In order to gain a clear understanding of the issues surrounding the platform workers' right to collective representation, the relevant key statutes and bills were referred to and carefully examined. The investigation also involved reviewing non-legal sources to uncover empirical data that was previously collected and reported, including in articles, journals, and reputable texts across various jurisdictions. The chapter begins with an exposition of conceptual frameworks pertaining to worker representation, highlighting its pivotal function in collective bargaining, the difficulties of adapting the notion to platform labour, and the rise of novel approaches to the concept. It proceeds by delineating the legislative frameworks and unionised activities of certain Asian nations' platform workers as they strive to improve their working conditions. Ultimately, the chapter concludes with a concise comparative analysis of collective representation regulatory frameworks and practices for platform workers in Asian countries, with brief references to the Western context. It deliberates on the potential advantages of collective power within the platform workforce. The next section examines the conceptual aspects of workers' entitlement to representation and the corresponding rights of collective bargaining, which serve as the foundation for this research work.

2.2 Key Concepts and Issues

2.2.1 The Origin of Worker Representation and Collective Bargaining

The notion of worker representation had its inception with the contributions of Sidney and Beatrice Webb (Dukes, 2008). Webbs believed an industrial constitution could reconcile capital and labour in the capitalist system (Dukes, 2008). The concept focuses on the expansion of trade unionism and the replacement of individual negotiations of terms and conditions with widespread collective bargaining, also known to be genuine freedom of contract. German lawyer Sinzheimer developed Webbs' industrial constitution notion and suggested state intervention to create trade unions and works councils. These bodies should have equal legal standing with employers in independent economic regulation (Dukes, 2008).

These ideas emerged during the second industrial revolution, which exposed the harsh working conditions endured by workers in large-scale production. These workers battled for improved working conditions (Dukes, 2008), where trade unions became the basis of collective worker organisations (Bertolini & Dukes, 2021) to forcefully advocate for the workers' interests and negotiate with employers for better working conditions (Freeman & Rogers, 2006). According to Md Radzi et al. (2022), a trade union serves as a strategic partner for employers by encouraging worker efficiency and professionalism. In contemporary labour law, International Labour Organisation Convention No. 98 (ILO, 1949) safeguards employees' rights to organise unions and participate in collective bargaining with employers.

2.2.2 *The Current State of Worker Representation and Collective Bargaining*

Over time, the significance of worker representation through trade unions and collective bargaining has been declining (Johnston & Land-Kazlauskas, 2018). The employment models of outsourcing, subcontracting, teleworking, and other techniques that cause displacement have played a role in this development (Weiss, 2011). In recent times, the rise of the platform economy, supported by the progress of digital technology, has caused additional disruption to the old system of worker representation (Schmalz et al., 2023). Regrettably, numerous legal systems worldwide continue to depend on the outdated paradigm of worker representation that grants rights only to individuals who precisely fit the characteristics of traditional employment arrangements.

Johnston and Land-Kazlauskas (2018), in their study of the challenges surrounding effective collective bargaining and representation for platform workers in the European and US contexts, reported that technological innovation and collective bargaining are often viewed as mutually exclusive; an inability to conceive of their coexistence is, in their words, merely a failure of imagination. The study primarily focused on the various approaches taken by European countries to worker participation in the platform economy. Due to technology improvements, it became evident that traditional negotiation strategies do not correspond with the new demands of workers in the platform economy.

Furthermore, Schmalz et al. (2023) found that the effectiveness of collective mobilisation strategies in the platform economy is significantly influenced by the different institutional, organisational, and structural contexts of individual countries. Their research showed how legal frameworks and union practices need to adapt to accommodate the distributed and often unstable nature of platform work. They concluded that the platform economy presents complex challenges that cannot be addressed through a one-size-fits-all approach.

Moreover, according to Weiss (2011), despite the growing complexity of workers' needs in a globalised, technologically driven labour market, the internationalisation of collective bargaining and worker representation is still a feasible goal. Weiss (2011) promoted a transnational approach to workers' rights, suggesting the creation of international frameworks to foster unity and strengthen cross-border negotiating power, especially for those working in the flexible but unpredictable environments of contemporary employment systems. The discussions that followed began with broad viewpoints and moved specifically to the Asian setting, establishing this point of view.

2.3 Discussion of Implications

2.3.1 Platform Workers' Challenges to Organising and Representing Collectively

A major challenge faced by platform workers in numerous countries is their exclusion from the purview of employment legislation, as they are typically categorised as independent contractors. Consequently, this classification impedes their ability to obtain union representation for the purpose of engaging in collective bargaining processes (Kerikmae & Kajander, 2022). Although platform workers are commonly categorised as independent contractors, platforms are progressively exerting influence over them while striving to keep the costs down (Kilhoffer et al., 2017). Essentially, the extensive control of platforms accurately reflected the characteristics of conventional employment, and numerous commentators emphasised this as a fundamental deception within a business model that essentially misclassified workers (Forsyth, 2023).

The features of platform work create additional challenges for collective representation. It has the characteristic of physically isolating and socially separating workers, which leads to the individualisation of platform workers and positions them as competing micro-enterprises rather than colleagues (Rafelis de Broves et al., 2024). Additionally, platform work involves a dispersed workforce and experiences a high rate of worker turnover (Kilhoffer et al., 2017). Platform employment often involves tasks that are of limited duration. The ability of platform workers to establish a community and find shared interests is undeniably affected by continually jumping between platforms, often without the intention of being on any platform for an extended period of time. These characteristics hinder efforts to achieve collective organisation (Johnston & Land-Kazlauskas, 2018).

Nevertheless, despite these obstacles, there is substantial evidence indicating that platform workers organise themselves collectively with the primary objective of advocating for their employment rights. The phenomenon occurred either through official means, such as when legislative bodies acknowledged platform workers' unions and allowed them to negotiate with platforms, or through informal methods, such as forming informal groups, establishing associations solely for their own benefit, or joining existing employee associations (Kilhoffer et al., 2017). Their collective actions, according to Forsyth (2023), consist of three strategies. Firstly, they engage in coordinated mobilisation to directly disrupt platform operations and advocate for increased regulation of platform work. Secondly, they employ strategic litigation to establish valuable legal precedents, garner public attention, and foster solidarity among workers with similar goals. Lastly, they negotiate collective agreements on behalf of platform workers. The authors provide several instances of the diverse representation approaches employed by platform workers in Western countries before transitioning to particular cases from Asian nations, as outlined in Table 2.1.

Additionally, it is noteworthy that there are converging patterns in the development of a collective entity among platform workers across Western countries and other regions. These patterns involve the creation of opportunities for platform workers to meet in person at certain sites inside the city. This promotes networking and socialising among riders as they utilise shared informal waiting places in urban centres, easily identifiable by their equipment (Rafelis de Broves et al., 2024). Platform workers enhance their online engagement by utilising various digital technologies and organisational strategies such as blog postings and Facebook groups. These

Table 2.1 Platform worker representation strategies and regulations in selected Western countries

Western countries	Platform worker representation strategies and regulations
France	Since 2016, platform workers in France have been legally entitled to participate in collective action, establish or join a union, and have their collective interests advocated for (Doherty & Franca, 2020)
UK	The UK labour laws recognise the entitlement of employees and workers to join trade unions and be represented by them in collective bargaining. Unfortunately, platform workers in the UK continue to struggle with the persistent problem of confusion over their job status (Doherty & Franca, 2020). This ambiguity presents a substantial obstacle, impeding the efficient use of trade unions as tools to campaign for the platform workers' rights However, platform workers in the UK have united under the umbrella of independent workers' unions to promote their interests. The Independent Workers Union of Great Britain (IWGB) and GMB Union were instrumental in pursuing a landmark case (*Uber BV v Aslam* [2021 UKSC 5) that resulted in the UK Supreme Court ruling that Uber drivers are considered workers, and hence entitled to the protections on minimum pay and working time established by the country. In addition, the IWGB has initiated legal action with the goal of securing the ability to engage in collective bargaining with Deliveroo on behalf of food delivery riders (*The Independent Workers Union of Great Britain v The Central Arbitration Committee and Roofoods Limited T/A Deliveroo* [2021] EWCA Civ 952). Despite its lack of success thus far, as Adams (2023) points out, unions have demonstrated their effectiveness in bolstering the negotiating power of both the union and its members when fighting for changes in public policy
Germany	The most significant advancement in the organisation of platform workers in Europe occurred in January 2016 when IG Metall, the largest union in Germany and the largest industrial union in the European Union, made the decision to allow self-employed workers to become members, specifically targeting crowd-and-platform-based workers. IG Metall successfully convinced eight online platforms to commit to a joint statement acknowledging their commitment to adhere to the minimum wage (Kilhoffer et al., 2017)

measures serve as a crucial initial stage in the process of establishing official unions, which are frequently accompanied by further face-to-face interactions (Aloisi, 2015).

2.3.2 *Asian Platform Workers' Union Movements and Regulations*

The majority of Asian legal authorities, similar to most Western countries, classify platform workers as independent contractors. Hence, like many Western countries, platform workers in Asia face the obstacle of relying on current legal frameworks that are typically associated with traditional employment relationships. This results in platform workers being denied the opportunity to organise unions and the right to be represented through collective bargaining. Nevertheless, recent developments have shown that the practices and regulatory framework for representing Asian platform workers have been effective in promoting collective representation.

2.3.2.1 Singapore

Self-employed individuals are the predominant perception of platform workers in Singapore as well. As a result, platform workers are not entitled for some legally mandated labour rights, which has led to the organisation of collective actions by these workers. Trenerry et al. (2022) documented the growing phenomenon of platform workers in Singapore coming together to mobilise as a collective force, extending their organising efforts beyond the usual boundaries of labour unions. An illustration of this phenomenon is the emergence of grassroots mobilisation by marginalised platform workers, such as the case of Grab Food Delivery Rider Singapore, where the number of delivery riders started to increase significantly in Singapore.

The Singapore government has made significant progress in enhancing the working conditions of platform workers by drafting a legislative framework to address their representation needs. As such, the proposal put forth by the Tripartite Workgroup on Representation for Platform Workers in Singapore (TWG) (TWG Report, 2023) is an insightful illustration for consideration when contemplating the legal framework governing the representation of platform workers. With the objectives of creating a representation framework that is specific to the requirements of platform workers and cultivating a strong industrial relationship between platform workers, their representative organisations, and digital platforms, the TWG was convened in August 2022. The new framework is considered crucial for creating channels through which platform workers may advocate for their interests and for digital platform firms to have clear and transparent mechanisms for resolving disputes.

The Singapore Government has acknowledged the recommendations from the TWG and undertook to establish a new legislative framework to enable representative

Table 2.2 Key recommendations of TWG (the Tripartite Workgroup on Representation for Platform Workers) on the representation framework for platform workers in Singapore

Areas	Key principles
Legal status of platform workers to unionise and collectively bargain	Platform worker organisations should possess the same level of authority as labour unions in order to negotiate legally binding collective agreements with platforms on behalf of workers
Mandate to represent platform workers	Platform worker representatives can get negotiating mandates through either direct recognition or secret ballots The establishment of a platform worker representative body is proposed when it receives the majority of votes cast, provided that at least 20% of eligible platform workers are present, and electronic voting is performed by the Ministry of Manpower (MOM)
Scope of negotiation	The TWG proposes that the legislation should refrain from defining the extent of discussions between representative organisations of platform workers and platforms, given the dynamic and evolving character of the business
Dispute resolution	A systematic protocol is suggested for resolving group conflicts, wherein the MOM acts as the first step for mediation prior to escalating the matter to the Industrial Arbitration Court
Overall objective	The focus of the representation framework is on ensuring fair treatment of platform workers while maintaining the flexible nature of platform labour

bodies to advocate for the interests of platform workers (Yufeng, 2023). The key recommendations reported by the TWG are summarised in Table 2.2.

To date (March 27, 2025), these recommendations have been enacted into law through the Platform Workers Act 2024 (No. 30 of 2024), which came into full force on 1 January 2025. By means of empowered representative bodies functioning similarly to trade unions, the over 88,000 platform workers in Singapore as of 2022 (approximately 3.6 percent of the resident labour force) now have the legal right to bargain collectively for improved working conditions (businesstimes.com.sg, 2023).

2.3.2.2 South Korea

In South Korea, it is anticipated that there are approximately 800,000 platform workers as of 2022, which accounts for 0.92% of the entire working population (Lee, 2023). Platform workers in South Korea have effectively mobilised collective representations, which serve as a dynamic and effective mechanism to address the needs of the groups they represent (Lee, 2023). This is despite the Labour Standards

Act and the Labour Union Act, which merely grant minimum employment rights to individuals classified as employees (including rights of association, collective bargaining, and collective action). By organising self-help meetings and protests, platform food delivery workers founded Rider Union in May 2019. Rider Union's core activists used rank-and-file organising based on their union advocacy experience. This grassroots union began a new campaign for safe rates to secure a minimum pay for platform workers, who must deliver more orders at a detrimental discount owing to algorithmic control (Rafelis de Broves et al., 2024).

The platform workers in South Korea indicated that they actively strive to have the right to engage in collective bargaining and be officially recognised as a union. In 2020, the Delivery Platform Union, a labour union associated with the Korean Federation of Service Workers' Union, actively participated in the Platform Labour Forum, a social dialogue for platform workers. This allowed them to establish institutional power and engage in collective negotiations with Baemin, Korea's largest food delivery company (Rafeli de Broves et al., 2024). In addition, various other associations were established to represent the diverse concerns of platform workers and promote solidarity (e.g., Life Magic Care Cooperative was established in 2018, and Korea Platform Freelancer Mutual Aid Association was established in 2021 (Jong-Jin, 2022)).

At the institutional level, the government has suggested modifications to current legislation to grant platform workers the rights to organise unions and engage in collective bargaining. According to the Labour Union Act, the term 'worker' encompasses individuals who are temporarily unemployed or actively seeking employment. Therefore, platform workers can be considered as workers under this Act. Nevertheless, the interpretation of the term 'employer' in the Act was construed to refer to the employer who has entered into a direct labour agreement with the workers. Therefore, the ability to form unions, negotiate collectively, and mobilise collectively is restricted to workers who are employed directly by the employer (Lee, 2023). There are several proposals for amending the legislation to facilitate the organised mobilisation of platform workers. Two essential proposals define a 'worker' as an individual who offers their labour to another person's work and receives compensation from the applicable business operator or labour recipient. On the other hand, the term 'employer' refers to an individual or entity that has actual influence or control over a worker's working conditions or performance. It can also refer to a person who acknowledges the other party's position in the trade union of the firm. The proposed definitions call for removal of the strict requirement for formal employment evidence, acknowledging that relationships involving platform and workers frequently do not adhere to formal contracts, yet still maintain a level of control over the workers involved. If the measure is approved, it has the potential to significantly expand the collective representation rights to platform workers (Lee, 2023).

2.3.2.3 Japan

In Japan, the focus of the topic over the collective representation of platform workers revolved around individuals employed in the food delivery industry rather than the ridesharing sector. The stagnation of the ridesharing industry in Japan can be attributed to the local law that prohibited its operation until recently (Uchiyama et al., 2022). Starting from April 1, 2024, the Japanese government partially removed restrictions on ride-hailing services, allowing private-hire automobiles to operate in four cities. This decision is aimed at addressing the scarcity of taxi drivers in the country (Koh, 2024).

UberEats became the focus of a case study to investigate the status and working conditions of platform workers since it operates in 27 Japanese cities with over 15,000 platform workers and holds a significant portion of the market for platform food delivery (as of 2019) (Umer, 2021). The escalating market for platform food delivery, which is projected to generate more than $6 billion in revenue by 2026, has given rise to a growing number of legal concerns concerning platform workers, most notably the deficiency of employment rights as they are regarded as self-employed (Uchiyama et al., 2022; Umer, 2021).

As a result, the UberEats Union was established in 2019 with the dual objectives of promoting collective bargaining to enhance working conditions for UberEats platform workers and establishing a legal framework that is nationally applicable to all platform workers (Uchiyama et al., 2022). The UberEats Union has taken actions by submitting a collective bargaining proposal to UberEats Japan, requesting fair wages and benefits and a review of the payment system. UberEats Japan refused to engage in negotiations, asserting that platform workers do not fall under the category of workers as defined by the Labour Union Act (LUA) (Ishibusi, 2022; Umer, 2021). The UberEats Union submitted a petition to the labour disputes arbitration department of the Tokyo Metropolitan Government Labour Relations Commission (LRC) and in November, 2022 UberEats Japan was requested to engage in collective bargaining by the LRC after the LRC acknowledged the riders' legal status as worker under the LUA. Article 3 of the LUA defines a 'worker' as an individual whose livelihood is dependent on wages, salaries, or other equivalent income.

In light of this triumphant legal proceeding, it is possible to assert that the UberEats Union has exemplified the efficacy of organising collective action, which commenced as non-legal union representation for alternative workers but progressed to institutional backing and sustained mobilisations.

2.3.2.4 Malaysia

According to the Department of Statistics Malaysia (DOSM) in 2020, there were 183,000 platform workers in Malaysia who were directly involved in ridesharing and food delivery services. As of 2021, 700,000 Malaysians were reportedly registered as ridesharing drivers or platform food delivery riders. This pattern demonstrated that the significant expansion of the Malaysian population relied on this industry for employment opportunities. Naharul (2022) exposes the fact that the general public was frequently deceived by data purporting that a ridesharing driver was making MYR7000 per month. Ultimately, the ridesharing driver found themselves working twelve-hour shifts for six consecutive days in order to generate a net income of only MYR4000, after deducting operating expenses. Platform workers were not afforded the protection of employment rights, the ability to form unions, or the right to engage in collective bargaining with digital platform companies under Malaysian law (Md Radzi et al., 2022; *Loh Guet Ching v Minister of Human Resources & Anor* [2022] 1 LNS 2388). The Industrial Relations Act of 1967 and the Trade Unions Act of 1959 serve as the primary legal basis for collective bargaining and the establishment of trade unions in Malaysia acknowledged the rights of conventional employees to form unions and participate in collective bargaining but are completely inapplicable to platform workers.

Thus, various associations serve as conduits for the platform workers' collective concerns. The recent controversy surrounding the low fares established by digital platforms has been contested by ridesharing drivers and delivery riders. The government was urged to set floor prices by representatives of the Grab Drivers Malaysia Association (GDMA), the Malaysian E-Hailing (Ridesharing) Association, and the Malaysian P-Hailing (Delivery) Riders Association (Penghantar), but this request was denied (Monihuldin, 2023; Nik Anis, 2023). Similarly, these associations collaborated with other non-governmental organisations, and activists advocated for the formation of the Malaysian Gig Economy Commission (SEGiM), an entity whose primary objective is to supervise the ecosystem of the gig economy in Malaysia (Adnan, 2023). SEGiM, which is composed of government officials, gig workers, and industry stakeholders is anticipated to contribute to the formulation and regulation of policies concerning the gig economy in Malaysia (Ministry of Human Resources, 2024).

There is a growing fear that the existing methods may not effectively accomplish the desired goals, despite the popularity of associations that seek to represent the interests of platform workers. Existing associations frequently lack formal mandates from workers and digital platforms, which presents a significant obstacle in the form of their limited functions (Md Radzi et al., 2022). In addition, a number of associations have arisen, purporting to represent certain categories of platform workers (Md Radzi et al., 2022). Nevertheless, the long-term sustainability of these entities is dubious due to their inability to function as labour unions, which includes the lack of legal power to obtain financial support.

It is worth noting that some prominent digital platforms choose to appoint appointed partners among the platform workers as representatives in order to obtain information regarding their working conditions (ILMIA, 2022). This demonstrates that platforms and platform workers can maintain healthy relationships through the use of effective channels for the collective voice of platform workers.

2.4 Comparison of Platform Workers' Collective Representation in Asian Countries

Table 2.3 below presents a summary of the different and coinciding legislative and collective representation practices observed in the selected Asian countries. Each Asian country has developed distinct regulatory strategies to manage the collective representation of platform workers. The representation of platform workers is perceived more positively in Singapore and South Korea, as the two countries are either progressing towards legislative recognition or have already implemented specific legal frameworks for the collective mobilisation and representation of platform workers. Japan has demonstrated a recognition of the representation rights of platform workers through judicial proceedings. However, Malaysia lacks both judicial and legislative frameworks that would grant its platform workers equivalent rights. While SEGiM has been established, its ability to fully support platform workers may be limited by its tripartite model. This structure is prone to various conflicts of interest, as each entity involved platforms, platform workers or their representative associations, and state authorities pursues distinct economic and political objectives. In the observed Asian countries, platform workers primarily relied on grassroots or alternative unions for collective mobilisation, as they were generally not recognised as eligible to form trade unions, including Singapore, which has only recently granted such rights.

2.5 Future Directions

The reliance of platform workers on collective mobilisation is commonly a result of their exclusion from the protection coverage of employment rights. Meanwhile, due to their classification as independent contractors, their ability to collectively organise and mobilise is not acknowledged under the laws. Therefore, platform workers in Asian countries, like certain Western countries, have endeavoured to depend on alternative sources of influence. They have utilised their ability to form associations, engage in discussions, negotiate in the workplace, and form alliances with various labour movement entities, ranging from work intermediaries to grassroots and established unions. Collective actions among platform workers are continuously evolving worldwide, especially in Asia, as they modify their plans and tactics to keep up with quickly changing technologies and managerial practices. Forsyth (2023) cautions that fully empowered collective representation demands that unions or grassroots associations of platform workers prioritise the advancement of workers' rights rather than competing for dominance or serving the interests of platform companies. Moreover, as platform work is intrinsically connected to broader social and economic dynamics, encouraging robust collective representation calls on platform worker groups to acknowledge the variety of identities present in their workforce, including marginalised people, for instance, women and migrant workers (Kim & Sawyer,

Table 2.3 Summary of the different and coinciding legislative and collective representation practices in the selected Asian countries

Country	Legislative safeguards for union formation, collective bargaining, and mobilisation rights	Platform workers' mobilisation channels	Significant successes in collective mobilisation among platform workers
South Korea	Bills were proposed to enable the platform worker representation	Alternative associations	No evidence
Japan	Negative	Alternative associations	Tokyo Labor Relations Commission's ruling on the status of UberEats Union for organising and collective bargaining under the Labor Union Act
Singapore	Platform Workers Act 2024 was enacted and fully came into force to legalise these rights	Alternative associations, but recently eligible to form legal associations, must satisfy the required conditions imposed by the Platform Workers Act 2024	No evidence
Malaysia	Negative	Alternative associations	Establishment of the Malaysian Gig Economy Commission (SEGiM)

2024; Lee, 2025; Rodríguez-Modroño et al., 2022). This implies that platform worker groups must not only prioritise the cultivation of solidarity but also critically evaluate and strategically determine which groups to align with, taking into account the unique socio-economic and regulatory contexts of individual Asian countries.

Furthermore, it has been seen that platform workers tend to align themselves with the institutional framework of employment, seeking recognition of their rights under worker representation laws. A variety of strategies have been employed to reach that juncture, such as initiating negotiations with platforms, securing the support of other organisations, enlisting the participation of authorities in the negotiation process, and pursuing strategic litigation. The utilisation of digital channels has made the process of mobilisation more efficient. For instance, utilising social media has proven to be highly beneficial in facilitating collaboration and, more significantly, in raising public consciousness regarding the importance of safeguarding employment rights. In spite of this, legal intervention was ultimately required to guarantee the efficacy of the collective actions, as direct collective bargaining with platform was deemed fruitless. This method signifies an adherence to the key idea of worker representation as established by Webbs. The support from the institutional structure is essential to ensure the success of the platform workers' collective actions (Duggan et al., 2024;

Piasna & Zwysen, 2022). In the context of UberEats Japan Union, an organisation that successfully mobilised its supporters through political campaigns and strategic litigation, ultimately obtained the right to unionise and negotiate collectively with platforms. As to Malaysia, it is evident that Malaysian platform workers operate with a comparable objective in terms of organising collective actions; however, they could benefit more from the assistance of grassroots unions in order to strategise their goals more effectively. Additionally, they can enhance their efforts by employing strategic litigation as a means of mobilisation, thereby increasing public awareness and political influence. Despite the potential cost burden, establishing a robust partnership with grassroots unions and other affiliated activist groups would provide valuable support in terms of legal representation and consultation. Regarding the legal framework, Singapore has provided a robust reference with its Platform Workers Act 2024, which is grounded in two essential concepts: inclusivity and flexibility. These concepts are particularly compatible with crafting a framework that addresses the unique characteristics of platform work. Concurrently, it continues to align with the current trajectory of labour regulation, which emphasises sustainability and inclusivity.

2.6 Conclusion

The surge in gig work, facilitated by digital platforms, has brought attention to the inadequacies of existing legal structures in safeguarding the rights and interests of platform workers. Scholars have a consensus that the traditional employment models and legal frameworks fail to encompass the dynamic and decentralised nature of platform work, leading to a discernible gap in representation and protection.

Hence, this study presents a framework and best practices for the representation of platform workers in the future of work in Asia, informed by insights from specific Asian jurisdictions. Nevertheless, this study has limits in its scope as it does not encompass a wider range of platform workers, such as crowdworkers, nor does it go into the practices and legal framework of other Asian nations. It also does not include first-hand empirical data to gain a more in-depth knowledge of issues concerning platform worker representation. Subsequent research must focus on bridging these gaps through first-hand empirical analysis and broadening the scope to include additional platform worker categories and jurisdictions.

In conclusion, the study demonstrated that the notion of worker representation remains highly pertinent to a contemporary form of employment facilitated primarily by digital platforms, provided that suitable strategies are implemented to ensure its efficacy in attaining its respective goals. Utilising effective mobilising of collective actions, supported by legislative provisions can be a highly effective strategy for establishing robust and productive relationships with platforms.

Highlights

- The notion of worker representation remains highly pertinent to a contemporary form of employment.
- Collective actions among Asian platform workers are continuously keeping up with quickly changing technologies and managerial practices.
- Legislative measures and flexible strategies strengthen platform worker representation in the future of work in Asia.

References

Adams, Z. (2023). Legal mobilisations, trade unions and radical social change: A case study of the IWGB. *Industrial Law Journal, 52*(3), 560–594.

Adnan, A. S. (2023, July 7). Penubuhan Suruhanjaya Ekonomi Gig Malaysia dijangka diumum dalam Belanjawan 2024. *Berita Harian.* https://www.bharian.com.my/berita/nasional/2023/07/1127750/penubuhan-suruhanjaya-ekonomi-gig-malaysia-dijangka-diumum-dalam

Aloisi, A. (2015). Commoditized workers. Case study research on labour law issues arising from a set of on-demand/gig economy' platforms. *Comparative Labor Law and Policy Journal, 37*(3), 653–690.

Araki, T. (2021). Collective bargaining for the self-employed in Japan. *Comparative Labor Law & Policy Journal, 42*(2), 449–464.

Bertolini, A., & Dukes, R. (2021). Trade unions and platform workers in the UK: Worker representation in the shadow of the law. *Industrial Law Journal, 50*(4), 662–688.

Cruz, S. A., & Gameiro, A. (2023). Digital work platform: Understanding platforms, workers, clients in a service relation. *Frontier in Sociology, 7*, 1075808. https://doi.org/10.3389/fsoc.2022.1075808

businesstimes.com.sg. (2023, August 14). Platform workers are spending more than they earn; savings in 'unhealthy range': DBS study. *The Business Times.* https://www.businesstimes.com.sg/singapore/platform-workers-are-spending-more-than-they-earn-savings-unhealthy-range-dbs-study

Department of Statistics Malaysia (DOSM). (2020). *Gig workers in Malaysia: A review of definitions and estimation.* DOSM Newsletter 8(1). Preprint retrieved from https://www.dosm.gov.my/v1/uploads/files/6_Newsletter/Newsletter%202020/DOSM_MBLS_1-2020_Series-8.pdf

De Stefano, V. (2015). The rise of the just-in-time workforce: On-demand work, crowdwork, and labor protection in the gig-economy. *Comparative Labor Law & Policy Journal, 37*, 461–471. https://doi.org/10.2139/ssrn.2682602

Doherty, M., & Franca, V. (2020). Solving the gig-saw? Collective rights and platform work. *Industrial Law Journal, 49*(3), 352–376.

Duggan, J., O'sullivan, M., & O'sullivan, M. (2024). Essential or excluded? Union pressures and state responses to platform work in three liberal market economies. *Transfer, 29*(4), 491–505. https://doi.org/10.1177/10242589241231731

Dukes, R. (2008). Constitutionalizing employment relations: Sinzheimer, Kahn-Freund, and the Role of Labor Law. *Journal of Law and Society, 35*(3), 341–363.

Durri, I. (2021). Regulating the platform economy: International perspectives on new forms of work. *Comparative Labor Law & Policy Journal, 42*(1), 237–244.

Forsyth, A. (2023). The digital resistance: Contesting the power of gig economy platforms through collective worker action. *Italian Labour Law e-Journal, 16*(2), 49–58. https://doi.org/10.6092/issn.1561-8048/18413

Freeman, R. B., & Rogers, J. (2006). *What workers want*. Cornell University Press.

Galpaya, H., Hurulle, G., Amarasinghe, T., & Bamaramannage, S. (2023, April). *Digital platforms in Asia: Access and use*. LIRNEasia Working Paper 2023/01. Retrieved March 8, 2023, from https://lirneasia.net/wp-content/uploads/2023/03/Digital-Platforms-Asia-LIRNEasia-WP2023_01-6Apr2023.pdf

Institut Maklumat dan Analisis Pasaran Buruh (ILMIA). (2022). *Kajian Mengenai Isu Perundangan Perburuhan, Perhubungan Perusahaan Dan Perlindungan Sosial*. ILMIA Jabatan Perangkaan Malaysia. Retrieved March 8, 2023, from https://www.ilmia.gov.my/index.php/en/component/zoo/item/article-kajian-gig-2022

International Labour Organisation (ILO). (1949). Right to organise and collective bargaining convention No. 98. Retrieved March 9, 2023, from https://normlex.ilo.org/dyn/nrmlx_en/f?p=NORMLEXPUB:12100:0::NO::P12100_INSTRUMENT_ID:312243

Ishibusi, K. (2022, July 21). For Japanese Uber delivery drivers, gig work is working. *Rest of World*. https://restofworld.org/2022/for-japanese-uber-delivery-drivers-gig-work-is-working/

Jong-Jin, K. (2022). *Growth of digital platform work in Korea and delay in the establishment of relevant legal institutions*. ILO EU KLI Conference: Policy dialogue on the future of work Presentation on Digital platforms and new technologies. Retrieved March 8, 2023, from https://www.ilo.org/sites/default/files/wcmsp5/groups/public/@ed_emp/documents/genericdocument/wcms_839426.pdf

Johnston, H., & Land-Kazlauskas, C. (2018). *Organizing on-demand: representation, voice and collective bargaining in the gig economy*. Conditions of work and employment series No. 94 ILO Office. Retrieved March 8, 2023, from https://www.ilo.org/publications/organizing-demand-representation-voice-and-collective-bargaining-gig

Kerikmae, T., & Kajander, A. (2022). Gig economy workers in the European Union: Towards change in their legal classification. *Revista CIDOB D'afers Internacionals, 131*(September), 117–136.

Kilhoffer, Z., Lenaerts, K., & Beblavý, M. (2017). *The platform economy and industrial relations applying the old framework to the new reality CEPS research report No. 2017/12*. Retrieved March 10, 2023, from https://aei.pitt.edu/88525/1/RR2017-12_PlatformEconomyAndIR_formatted.pdf

Kim, P., & Sawyer, S. (2024). Occupational diversity in platform work: A comparative study. In *Companion of the 2024 computer-supported cooperative work and social computing (CSCW Companion '24)* (pp. 209–215). ACM. https://doi.org/10.1145/3678884.3681853

Koh, S. (2024, April 1). Japan lifts partial ban to give green light to ride-hailing services in Tokyo, Kyoto *The Star*. https://www.thestar.com.my/aseanplus/aseanplus-news/2024/04/01/japan-lifts-partial-ban-to-give-green-light-to-ride-hailing-services-in-tokyo-kyoto

Lee, S. S. (2023, March). *Desk study EU-Korea agreement platform work and institutional protection with a specific focus on South Korea and the EU. Support to civil society participation in the implementation of EU trade agreements EuropeAid/139031/DH/SER/MULTI*. Retrieved April 10, 2023, from https://www.eesc.europa.eu/sites/default/files/files/dest_study_eu-korea_fta_platformwork_lee2023april15.pdf

Lee, Y. (2025). Class identity vs intersectional solidarities: Divergent models for organizing gig workers in Seoul and Toronto. *International Journal of Comparative Sociology*. https://doi.org/10.1177/00207152241312904

Md Radzi, M. S. N., Bidin, A., Musa, M. K., & Hamid, N. A. (2022). Protecting gig workers' interests in Malaysia through registered association under Societies Act 1966. *IIUM Law Journal, 30*(SI), 157–179.

Ministry of Human Resources. (2024, July 4). Mesyuarat Jawatankuasa High Level Committee (HLC) Penubuhan Suruhanjaya Ekonomi Gig Malaysia Bil. 1/2024, https://www.mohr.gov.my/images/pdf/KENYATAAN%20MEDIA%204.7.2024%20MESYUARAT%20JAWATANKUASA%20HIGH%20LEVEL%20COMMITTEE%20(HLC)%20PENUBUHAN%20SURUHANJAYA%20EKONOMI%20GIG%20MALAYSIA.pdf

Monihuldin, M. (2023, November 17). Ehailing drivers floored by 'unreasonable' fares. *The Star*. https://www.thestar.com.my/news/nation/2023/11/17/ehailing-drivers-floored-by-unreasonable-fares

Naharul, MA. (2022, October 18). The gig economy: What monster are we breeding? *The Star*. https://themalaysianreserve.com/2022/10/18/the-gig-economy-what-monster-are-we-breeding/

Nik Anis, M. (2023, April 11). Transport Ministry prefers to let supply and demand set ehailing fares *The Star*. https://www.thestar.com.my/news/nation/2023/04/11/ehailing-services-request-for-floor-price-being-considered-says-transport-minister

Piasna, A., & Zwysen, W. (2022). New wine in old bottles: Organizing and collective bargaining in the platform economy. *International Journal of Labour Research, 11*(1–2)

Rafélis de Broves, O., Kang, M., Grohmann, R., Barcellos, B., Gomes Mano, F., Yoon, C., & Jamil, R. (2024). Workers organizing in the platform economy: Local forms and global trends of collective action. *Sociology Compass, 18*(2), Article e1318. https://doi.org/10.1111/soc4.13188

Rodríguez-Modroño, P., Agenjo-Calderón, A., & López-Igual, P. (2022). Platform work in the domestic and home care sector: New mechanisms of invisibility and exploitation of women migrant workers. *Gender & Development, 30*(3), 619–635. https://doi.org/10.1080/13552074.2022.2121060

Schmalz, S., Basualdo, V., Serrano, M., Vandaele, K., & Webster, E. (2023). Varieties of platform unionism: a view from the Global South on workers' power in the digital economy. *Work in the Global Economy, 3*(2), 201–224. Retrieved March 28, 2025, from https://doi.org/10.1332/27324176Y2023D000000001

Trenerry, B., Chng, S., Wang, Y., Sun Lim, S., & Ho Oh, P. (2022). Mobilising workers via digital means: Reinventing trade unions for the digital age. *Singapore Labour Journal, 1*(1), 6–19. https://doi.org/10.1142/S2811031523000025

Tripartite Workgroup on Representation for Platform Workers (TWG). (2022, July 11). *Enhancing representation for platform workers*. Ministry of Manpower Singapore. Retrieved March 8, 2023, from https://www.mom.gov.sg/-/media/mom/documents/employment-practices/platform-workers/twg-on-representation-for-platform-workers-report.pdf

Uchiyama, Y., Furuoka, F., Md Akhir, M. N., Jingyi, L., Beatrice, L. & Hanim Pazim, K. (2022). Labour union's challenges for improving for gig work conditions on food delivery in Japan: A lesson for Malaysia. *Wilayah The International Journal of East Asian Studies, 11*(1), 83–111. https://doi.org/10.22452/IJEAS.vol11no1.7

Umer, H. (2021). Illusory freedom of physical platform workers: Insights from Uber Eats in Japan. *The Economic and Labour Relations Review, 32*(3), 437–452. https://doi.org/10.1177/1035304621992466

Umney, C., Stuart, M., Bessa, I., Joyce, S., Neumann, D., & Trappmann, V. (2024). Platform labour unrest in a global perspective: How, where and why do platform workers protest? *Work, Employment and Society, 38*(1), 3–26. https://doi.org/10.1177/09500170231209676

Weiss, M. (2011). Re-inventing labour law? In G. Davidov & B. Langille (Eds.), *The idea of labour law* (pp. 43–54). Open University Press.

Yang, K., & Joo Chae, Y. (2020). Organizing the young precariat in South Korea: A case study of the Youth Community Union. *Journal of Industrial Relations, 62*(1), 58–80. https://doi.org/10.1177/00221856198682

Yufeng, K. (2023, July 14). Platform workers in Singapore to get more bargaining power, better representation *The Straits Times*. https://www.straitstimes.com/singapore/platform-workers-in-singapore-to-get-more-bargaining-power-better-representation

Nazruzila Razniza Mohd Nadzri is a Senior Lecturer at the Faculty of Business and Management, Open University Malaysia. She holds a Doctoral of Philosophy (Ph.D.) in Law from the National University of Malaysia. Dr. Nazruzila Razniza is not only a teaching staff but an active researcher with numerous publications in e-journals, journals, magazines and conference proceedings. Her research interests encompass various aspects of employment law, contemporary employment relations, and recent legal issues relating to workers' rights and the changing nature of work. Her passion for scholarly work continues to make significant contributions to the field of law and policies.

Nusrah Samat is a committed researcher who has a significant interest in management techniques and behavioural research. Her work has significantly advanced these subjects by illuminating the complexities of organisational human behaviour. Dr. Nusrah Samat is known for her knowledge of behavioural research and her insights into management practices, particularly in the areas of total quality management, market orientation, and service quality. Her research enhances the theoretical understanding of these areas and provides practical insights that benefit both businesses and academia. Her commitment to excellence and innovative approach established her contribution in the fields of behavioral studies and management practices.

Sharmila Devi is currently holding the position of Senior Lecturer at Inti International University. With diverse 20-year tenure in the academic realm, her responsibilities have encompassed guiding students, spanning across undergraduate and postgraduate levels, and catering to both local and international cohorts. As an expert in Human Resource Development, Dr. Sharmila Devi meticulously developed an array of modules, assessments, and learning resources tailored to cater to the needs of learners. Her passion for HRM (Human Resource Management) has also driven her to explore various facets of this discipline, including Green HRM, Women Leadership, Spiritual Intelligence, Leadership Development, and Talent Management.

Chapter 3
Shaping the Future of Work: Leveraging Environmental, Social, and Governance (ESG) as a Strategic Human Resource Strategy

Cluster of Future Studies: Corporate Foresight

Jing Yi Yong, Woon Leong Lin, Thanuja Rathakrishnan, Thivashini B. Jaya Kumar, Feranita Feranita, and Nelvin XeChung Leow

Abstract This conceptual chapter delves into the transformative potential of integrating environmental, social, and governance (ESG) principles into human resources (HR) strategy for shaping the future of work. With increasing environmental concerns, societal expectations, and governance reforms in Asia, ESG metrics have become critical for investors and stakeholders. Proactively incorporating ESG principles into HR strategies allows organisations to navigate regulatory changes, enhance resilience, and align with contemporary values. The chapter explores how ESG integration impacts talent acquisition, retention, employee engagement, and corporate culture while addressing challenges in implementing ESG-focused HR strategies. It also reviews methods for measuring ESG's influence on HR outcomes,

J. Y. Yong (✉) · W. L. Lin · T. Rathakrishnan · T. B. Jaya Kumar · N. X. Leow
School of Management and Marketing, Faculty of Business and Law, Taylor's University, Subang Jaya, Malaysia
e-mail: serene.yong@taylors.edu.my

W. L. Lin
e-mail: woonleong.lin@taylors.edu.my

T. Rathakrishnan
e-mail: thanuja.rathkrishnan@taylors.edu.my

T. B. Jaya Kumar
e-mail: thivashini.jk@taylors.edu.my

N. X. Leow
e-mail: nelvinxechung.leow@taylors.edu.my

F. Feranita
Nottingham University Business School China, University of Nottingham Ningbo China, Ningbo, China
e-mail: feranita.feranita@nottingham.edu.cn

C. Y. Fung et al. (eds.), *Future of Work in Asia*,
https://doi.org/10.1007/978-3-031-95804-5_3

providing insights into how these practices can drive sustainable business success. By synthesising concepts from organisational behaviour, human resource management, sustainability studies, and strategic management, this chapter offers a robust theoretical framework for ESG-driven HR strategies. Through this lens, it highlights the strategic importance of aligning HR practices with ESG principles to create a sustainable, engaged, and future-ready workforce. This chapter serves as a roadmap for HR professionals and organisations, offering practical guidance to navigate the complexities of contemporary business environments while meeting societal and environmental expectations.

Keywords Environmental · Social · and Governance (ESG) · Human resource strategy · Talent acquisition · Talent retention · Employee engagement · Corporate culture

3.1 Introduction

The dynamics of the workplace are experiencing significant shifts, driven by the forces of technology, globalisation, and evolving social values. The work environment and the very nature of work exert considerable influence on critical aspects such as productivity, job satisfaction, health, safety, and overall wellbeing, all intricately tied to the concept of sustainable development (Torres et al., 2023). In light of these transformations, organisations are compelled to adapt to stay competitive and cultivate an environment that not only attracts but also retains top talent. Recognising the indispensable role in this adaptive process, organisations are progressively embracing the integration of Environmental, Social, and Governance (ESG) principles into their operational frameworks. It is essential to underscore that the United Nations Principles for Responsible Investment delineate ESG across three dimensions, as illustrated in Table 3.1.

The Asian region is currently experiencing a heightened focus on sustainable and ethical business practices, driven by environmental considerations, societal expectations, and governance reforms. Notably, investors and stakeholders in Asian markets are increasingly acknowledging the importance of ESG metrics, recognising their

Table 3.1 Examples of ESG (Environmental, Social, and Governance) factors (reproduced from United Nations Global Compact, 2021)

Environment	Social	Governance
• Climate change • Resource depletion • Waste • Pollution • Deforestation	• Working conditions • Health and safety • Relationships with employees • Human rights	• Bribery and corruption • Board diversity and structure • Executive pay • Tax strategy

role in promoting sustainable business practices (Ng et al., 2020). Concurrently, regulatory developments in certain Asian markets have been instrumental in advancing the ESG and sustainability agenda. For instance, Malaysia introduced the Sustainable and Responsible Investment (SRI) and ESG reporting guide, while China initiated the Green Investment Principles for the Belt and Road and implemented ESG disclosure requirements for listed companies. Similarly, Singapore has published the Sustainability Reporting Guide and the Singapore Stewardship Principles for Responsible Investors in 2016, among other initiatives (PwC, 2020).

ESG has ascended to a prominent position on the boardroom agenda, and its significance goes beyond capturing the attention of the investor community concerned with its vital role in guaranteeing business sustainability. It is transforming from a framework for sustainable and responsible business practices into a dynamic force shaping the future of work. This transformation is clearly manifest in the intersection of ESG considerations with Human Resource (HR) strategies, notably within the domain of employee engagement. The 2022 Edelman Trust Barometer (Edelman, 2022) highlights this intersection, revealing that 60% of over 36,000 global respondents base their choice of workplace on beliefs and values. Core ESG topics such as culture, climate change, business ethics, and Diversity, Equity, and Inclusion (DE&I) have become pivotal factors influencing both current employees and potential candidates. The alignment of organisational values with these ESG elements has become a significant determinant in attracting and retaining top talent in the competitive landscape of the modern workforce.

In the Asian business landscape, sustainable practices have become integral to corporate strategies. Employers are actively seeking individuals capable of leading transformative change in this context. Kerry Consulting, an executive recruitment firm, has responded by launching a dedicated ESG recruitment practice, emphasising the recruitment of top-tier talent for sustainability initiatives and change programs in Asia (Liew, 2023). The influence of ESG extends beyond recruitment to shaping the employee experience in organisations. Aligning ESG principles with employee values enhances satisfaction, engagement, and wellbeing, resulting in improved talent retention. A notable example is Singapore-based DBS bank, where a commitment to sustainability and the integration of ESG values in the workplace provide a compelling competitive advantage, particularly in talent retention within the competitive labour market (Tan, 2023).

Given the social expectations and regulatory changes regarding ESG compliance in many Asian countries, the proactive integration of these principles into HR strategies positions companies to address evolving legal standards, fostering a resilient and forward-looking organisational framework. Utilising a methodological framework that integrates literature review and real-world case studies, our objective is to meticulously analyse and interpret the concepts presented. This multifaceted approach aims to provide a comprehensive understanding of the implications, effectiveness, and hurdles associated with the incorporation of ESG principles within HR strategies. By synthesising theoretical foundations with empirical evidence, the chapter offers actionable insights and practical guidance for organisations and HR professionals to align HR strategies with ESG principles, thereby catalysing sustainable business success and fostering societal impact. Consequently, this chapter pursues threefold objectives: (i) to discover the intersection of ESG principles with HR strategies, specifically on talent acquisition, talent retention, employee engagement, and overall corporate culture, (ii) to explore methods for measuring the influence of ESG on HR outcomes, and (iii) to delve into the challenges and considerations inherent in implementing an ESG-focused HR strategy.

3.2 Key Concepts and Issues

3.2.1 The Intersection of Environmental, Social, and Governance (ESG) Principles and HR Strategies

ESG has solidified its position as a foundational value intricately woven into business strategy (Jasni et al., 2020). As ESG gains increasing prominence in the broader business landscape and its role extends to shaping the future workforce, our study delves into a novel intersection. Specifically, we explore how ESG principles intersect with various HR functions, encompassing talent acquisition, talent retention, employee engagement, and corporate culture. This inquiry seeks to uncover the nuanced connections between ESG values and key aspects of human resource management, recognising the evolving dynamics where responsible business practices deeply influence the overall employee experience and organisational culture.

3.2.1.1 Talent Acquisition

The trajectory of the future of work is evolving rapidly, guided by the perspectives of Millennials and Gen Zs. Embedding ESG principles into talent acquisition involves aligning organisational values with prospective employees committed to sustainability and responsible business practices. In a 2021 survey conducted by international real estate experts JLL in Singapore, involving 200 Singaporean employees, unveil a crucial trend: the younger generations prioritise ESG factors as the purpose of business. Over 50% would choose a socially and environmentally responsible employer over similar offers. Moreover, 67% prefer companies that involve employees in sustainability, with this preference rising to 70–80% among those aged 21–30. This signals a fundamental shift, with companies excelling in ESG having a competitive edge in attracting new talent (Tan, 2022).

Furthermore, the integration of ESG principles not only enhances an organisation's appeal to potential employees but also suggests that high ESG scores can confer a competitive advantage in talent acquisition, thereby augmenting corporate value. Acknowledging the growing significance of corporate social responsibility (CSR), organisations are increasingly recognising the necessity to communicate their commitment to employees, the community, and other stakeholder groups as part of their efforts to attract applicants (Jakob et al., 2022). As highlighted by Karwowski and Raulinajtys-Grzybek (2021), organisations that report their sustainability practices tend to enjoy more positive reputations, thereby increasing their likelihood of attracting and retaining top talents. This perspective is further supported by Mohammad and Wasiuzzaman (2021), who contend that in emerging markets like Malaysia, creating value through the strategic integration of ESG principles, coupled with a clear vision, is key to attracting the best talent.

3.2.1.2 Talent Retention

In a continuously changing business environment, HR plays a vital role in securing and retaining essential talent for organisational success. ESG principles are key to fostering purpose and aligning with individual values. Research by Stahl et al. (2020) indicates that organisations characterised by a clear sense of purpose, with a focus on prioritising employees as stakeholders, tend to exhibit higher levels of alignment between organisational and employee goals, culminating in a more motivated workforce. In essence, organisations that emphasise sustainability, ethical governance, and social responsibility are strategically positioned to retain talent over the long term. This is supported by Zainee and Puteh (2020), showing that all aspects of CSR, including, economic, legal, ethical, and philanthropic responsibilities, positively influence the retention of Gen Y accountants in Malaysia. The authors assert that organisations aligning with these CSR principles boosts emotional attachment, attracts talent, and fosters loyalty, reflecting Gen Y's preference for socially responsible employers.

Moreover, employees are more inclined to remain with organisations that offer a sense of purpose extending beyond mere financial objectives—an aspect inherently delivered by ESG initiatives. A 2022 Mojodomo survey of 1097 employees across Asia reveals that 91% of respondents think companies should prioritise all stakeholders, not just shareholders. Moreover, 92% align their personal beliefs with ESG values, and a striking 98% expect their employers to make a positive global impact (Mojodomo, 2022). ESG-focused organisations, often perceived as forward-thinking and responsible (Asif et al., 2023), demonstrate a commitment to investing in employee development (Nekhili et al., 2021). By offering opportunities for employees to actively engage with ESG activities, particularly social initiatives, organisations not only contribute to professional growth but also enhance job satisfaction, leading to increased productivity and ultimately resulting in higher retention rates (Piao et al., 2022).

3.2.1.3 Employee Engagement

Employee engagement thrives in environments that prioritise ESG principles. The integration of ESG not only imparts a sense of purpose but also fosters a belief in contributing to broader societal and environmental objectives, thereby amplifying employee engagement. Asian employees increasingly value happiness and purpose over salary alone, reflecting a shift towards more meaningful work. This trend has been amplified by the pandemic, emphasising the importance of ESG initiatives (Mojodomo, 2022). ESG initiatives, ranging from volunteer programs to environmental awareness campaigns and support for social causes, instil a deep sense of pride and commitment among the workforces. A survey by Mojodomo (2022) indicates that 69% of Asian employees express interest in ESG programs, with mental health (76%), environmental sustainability (68%), and community involvement (58%) topping the interest list.

According to Malhotra and Pachauri (2023), ESG initiatives signify a heightened concern for employee wellbeing, contributing to positive employer branding and a favourable positioning in the labour market, consequently increasing employee engagement. Furthermore, ESG initiatives frequently involve employees in impactful projects related to sustainability or community engagement. The favourable reception of employees toward CSR/ESG initiatives enhances engagement, motivation, and commitment levels, playing a fundamental role in their dedication to the organisation and corporate performance. A high level of CSR/ESG in an organisation promotes a better reputation, encouraging integrity among employees and leads to improved employee engagement (Tunio et al., 2021). In other words, an organisation that creates a positive and fulfilling work experience can motivate employees to give their best, thereby increasing their engagement with the organisation.

3.2.1.4 Corporate Culture

ESG principles are instrumental in shaping a positive corporate culture. The integration of ESG reinforces a culture that prioritises values associated with sustainability, ethics, and social responsibility, establishing a foundation for a culture centred on accountability and shared responsibility (Atkins et al., 2023; Kotsantonis & Serafeim, 2019). In Asia, the ESG landscape has undergone a remarkable evolution, with over 80% of listed companies embracing ESG metrics in their reporting by 2023, a significant rise from a mere 5% a decade earlier. This momentum isn't limited to large corporations; startups and SMEs are also embracing responsible practices, highlighting the integral link between ESG efforts and organisational culture. This reflects the increasing recognition that sustainability is not just a business strategy, but a fundamental value embraced by companies across various sizes and sectors in Asia (Saluto, 2024).

ESG, often inclusive of diversity and inclusion initiatives, provides an opportunity for human resource development practitioners to act as change agents, aiding organisations in cultivating socially responsible and ethical behaviours, and transforming organisational culture (Cho et al., 2021). A culture that embraces ESG values not only fosters innovative thinking and problem-solving but also drives positive change (Foster et al., 2023). Additionally, Zahari et al. (2024) underscores the significance of ethical leadership in reducing integrity violations, particularly in the realm of ESG compliance. Ethical leaders contribute to a culture where adherence to ESG principles is prioritised. In view of this, HR professionals and leaders play a crucial role in championing and embedding these principles throughout the organisation, shaping a workplace culture that not only attracts and retains talent but also motivates employees.

3.2.2 *ESG Metrics: Evaluating the Impact on Human Resources Performance*

The incorporation of ESG principles into HR strategy yields significant benefits; however, assessing their influence on various HR aspects is crucial. This includes talent acquisition, retention, and employee engagement, as well as shaping corporate culture. The process employs the development of specific ESG-related metrics and Key Performance Indicators (KPI) to track progress and impact on HR outcomes. Regular review and analyses of these metrics facilitate strategy adjustments, ensuring alignment with organisational goals and providing valuable insights for overall success.

3.2.2.1 Talent Acquisition

Research indicates that organisations prioritising ESG principles tend to enhance their employer reputation positively (Jakob et al., 2022; Karwowski & Raulinajtys-Grzybek, 2021). To gauge the impact, it is crucial to assess shifts in the organisation's reputation as an employer renowned for ESG commitment. A strong employer branding orientation contributes to building a distinctive and appealing external image, potentially improving recruitment efficiency in terms of time and cost. Elevated employer branding orientation enhances recruitment efficiency by attracting a pool of applicants aligning with the organisation's culture and values. This results in reduced time and costs needed to fill positions appropriately (Tumasjan et al., 2020). Table 3.2 outlines key ESG metrics and corresponding evaluation methods to assess the attractiveness of the organisation to top talent.

3.2.2.2 Talent Retention

Addressing high turnover rates and associated replacement costs is a significant workforce challenge (Johnson et al., 2021). ESG principles are believed to instil a profound sense of purpose and alignment with individual values, enhancing employee commitment. Therefore, it is imperative to analyse whether ESG-focused HR strategies reduce turnover rates and assess whether employees who resonate with ESG principles are more likely to stay with the organisation. Additionally, as employees favour organisations offering learning and development opportunities (Nekhili et al.,

Table 3.2 Evaluation of ESG (Environmental, Social, and Governance) on talent acquisition

Evaluation category	Methods	Key metrics
Employer reputation and branding	Surveys	Conduct surveys among current employees, job applicants, and new hires to assess their perception of the organisation's ESG commitment and its impact on the employer's reputation
	Online presence	Monitor online platforms, including employer review sites like Glassdoor or Indeed, to gauge public perception of the organisation's ESG initiatives and how they influence the employer brand
Recruitment efficiency	Time-to-fill analysis	Measure the average time taken to fill job vacancies. A decrease in time-to-fill for job vacancies can indicate increased attractiveness and alignment of top talent with the organisation's ESG values
	Cost analysis	Compare pre- and post-ESG initiative recruitment costs. A decrease in recruitment costs post-implementation can signify improved efficiency in attracting and hiring suitable candidates

2021; Piao et al., 2022), assessing the impact of ESG-related training programs on professional growth is essential. Table 3.3 outlines key evaluation methods for assessing the impact of ESG initiatives on talent retention.

Table 3.3 Evaluation of ESG (Environmental, Social, and Governance) on talent retention

Evaluation category	Methods	Key metrics
Turnover rates	Historical analysis	Compare turnover rates before and after implementing ESG-focused HR strategies to assess any significant changes or improvements
	Benchmarking	Benchmark turnover rates against industry standards to evaluate the organisation's performance in talent retention with ESG initiatives in place
Employee training and development	Training participation rates	Measure participation rates in ESG-related training programs to assess employee engagement and interest in professional development opportunities
	Skill development	Evaluate the development of new skills and competencies among employees who have participated in ESG-related training programs to understand their impact on career growth and retention

Table 3.4 Evaluation of ESG (Environmental, Social, and Governance) on employee engagement

Evaluation category	Methods	Key metrics
Employee engagement	Employee feedback	Use surveys to gather feedback from employees about their perceptions of the company's ESG initiatives and how they impact their engagement and motivation
	Focus group	Conduct focus group discussions to delve deeper into employees' opinions, experiences, and suggestions regarding ESG factors and engagement
Job satisfaction and productivity	Employee surveys	Use surveys to gauge changes in job satisfaction and productivity levels following the implementation of ESG initiatives
Value alignment	Employee surveys	Assess the alignment of ESG values with employee expectations and perceptions

3.2.2.3 Employee Engagement

To comprehensively evaluate the impact of ESG initiatives on HR outcomes, it is crucial to adopt a multifaceted approach. Saks (2019) suggests evaluating employee engagement through key indicators such as job satisfaction, organisational commitment, and intention to quit. Recognising the varied perspectives among academicians and practitioners on the construct of employee engagement, Shrotryia and Dhanda (2019) developed a comprehensive instrument covering three dimensions: alignment, affective, and action-oriented. This scale encompasses value alignment, empowerment, a sense of commitment and belongingness, and contribution to work, providing a holistic measurement of employee engagement. Table 3.4 presents the methods used to measure the effect of ESG on employee engagement.

3.2.2.4 Corporate Culture

The incorporation of ESG principles into the corporate culture can be assessed by examining how these values permeate day-to-day operations and decision-making processes. To measure changes in organisational culture resulting from ESG integration, a cultural survey can be employed. Organisations are progressively embracing initiatives to incorporate ESG activities into their decision-making processes, placing emphasis on non-financial reporting and aligning performance with Sustainable Development Goals (SDGs). This strategic approach, advocated by Bukreeva and Grishunin (2023), underlines the importance of communicating these efforts to stakeholders. Qualitative feedback from stakeholders, including employees, customers, and suppliers, can provide insights into cultural shifts. An ESG framework committed to social and environmental performance, such as reducing carbon footprint, can be integrated with participatory and inclusive governance practices, including workplace culture and a commitment to diversity and inclusion, as highlighted by Tavanti (2023). As a result, surveys or qualitative assessments can be utilised to determine if

Table 3.5 Evaluation of ESG (Environmental, Social, and Governance) on corporate culture

Evaluation category	Methods	Key metrics
ESG integration assessment	Cultural survey	Implement cultural surveys to assess the extent to which ESG principles have been incorporated into day-to-day operations and decision-making processes
	Workplace culture evaluation	Measure employees' perceptions of workplace culture, focusing on inclusivity, equity, and opportunities for growth and development
Stakeholder feedback	Qualitative feedback collection	Gather qualitative feedback from stakeholders, including employees, customers, and suppliers, to gain insights into cultural shifts resulting from ESG integration

employees from diverse backgrounds feel included and valued. Table 3.5 shows key evaluation methods for assessing the impact of ESG initiatives on corporate culture.

3.2.3 Navigating the ESG Landscape: Challenges and Considerations in Implementing HR Strategies

While ESG-focused HR strategies hold significant potential for shaping the future of work, their implementation poses inherent challenges. This is especially notable in Asia, where ESG is a relatively novel concept, and many industry players are just commencing their exploration of ESG-compliant frameworks to embed these values in their business operations. Navigating these challenges effectively becomes crucial for successful integration. In this section, we delineate three major challenges: cultural awareness and adaptation, limited understanding of the ESG framework, and balancing short-term and long-term objectives.

3.2.3.1 Cultural Awareness and Adaptation

Asia, home to over 4 billion people, showcases a rich tapestry of cultural diversity that leads to varied attitudes toward sustainability and social responsibility. While some employees in the region readily align with ESG principles that resonate with their cultural values, others resist what they perceive as foreign or conflicting with tradition. Ling's (2019) comparative study across Taiwan, Singapore, and Malaysia underscores the profound impact of cultural and contextual factors on CSR practices. The study reveals that managerial cultural values, notably Confucianism (with its emphasis on long-term orientation) and collectivism, significantly shape CSR initiatives. Managers embracing these values prioritise CSR, particularly focusing on employee welfare and environmental sustainability.

Country-specific factors further influence CSR strategies, with power distance, uncertainty avoidance, and the context of origin and operation playing pivotal roles. Taiwanese companies, deeply rooted in their cultural ethos and societal expectations, exhibit a stronger commitment to CSR compared to their Singaporean and Malaysian counterparts (Ling, 2019). However, resistance to ESG adoption can be encountered, particularly in regions where ESG concepts are nascent. A limited awareness of ESG benefits and the introduction of new management practices driven by the ESG era can induce stress and resistance among both employees and management. Therefore, successfully adapting ESG principles across Asia's diverse cultural landscape requires a nuanced understanding of local values and principles (Poppelwell & Overton, 2022).

3.2.3.2 Limited ESG Framework Understanding

Various global frameworks for ESG reporting, such as the Sustainability Accounting Standards Board (SASB), the Global Reporting Initiative (GRI), Integrated Reporting (IR), and Task Force on Climate-Related Financial Disclosures (TCFD), have been established. These frameworks aim to guide organisations in reporting their ESG performance, ensuring transparency and accountability to stakeholders. However, ESG adoption is still in its nascent stages across many Asian regions, resulting in a limited understanding of the comprehensive ESG framework among businesses and HR professionals. The lack of clarity on regulatory and reporting standards hinders widespread ESG integration and implementation (Parameswar et al., 2024). In terms of ESG disclosures, Singhania and Saini (2022) categorised countries like Singapore, India, China, the Philippines, Malaysia, and Argentina as having a developing-stage ESG framework, while Russia, Indonesia, Thailand, Nigeria, and Vietnam are classified as having an early-stage framework due to low ESG scores.

The regulatory framework for ESG practices varies significantly across Asian countries. Singapore mandates climate reporting for listed companies, Hong Kong SAR requires listed firms to include ESG-related risks in their enterprise risk management assessments, and Japan's revised Corporate Governance Code mandates TCFD compliance for Prime Market listed companies. This diversity presents challenges for multinational corporations operating across regions like China, the US, UK, and Europe, requiring them to navigate different reporting requirements (PwC, 2023). The lack of consistent company disclosures further complicates the adoption of ESG-focused HR strategies. Without standardised reporting and clear regulatory guidelines, businesses struggle to benchmark, measure, and improve their ESG performance. This variability in ESG maturity and regulatory frameworks underscores the complexity of ESG adoption in Asia, emphasising the need for tailored strategies to foster sustainable business practices regionally. Developing ESG capabilities and upskilling employees become crucial to keeping them abreast of evolving local and global sustainability reporting standards.

3.2.3.3 Balancing Short-Term and Long-Term Objectives

Organisations often grapple with the challenge of delivering immediate results and meeting short-term objectives, a paradigm that may clash with the inherently longer-term nature of ESG initiatives. Asia's rapid economic growth and diverse cultural landscape add layers of complexity to this balancing act. The Asian business environment is often characterised by intense competition due to ASEAN free trade agreements (Mihardjo et al., 2020) and a strong emphasis on quarterly performance (Kim, 2022). This short-term focus is reinforced by investors and shareholders who prioritise immediate financial returns, sometimes at the expense of ESG considerations. Schoenmaker and Schramade (2019) assert that prevailing business practices exhibit a narrow focus on short-term financial returns and attempts at ESG integration tend to be insufficiently deep to address this issue.

Balancing the need for short-term outcomes with long-term ESG goals within an HR strategy is intricate, given the imperative to satisfy stakeholders' immediate expectations while concurrently investing in sustainable practices for enduring benefits. For organisations guided by purpose-driven strategies, finding equilibrium between fulfilling stakeholders' needs, meeting short-term financial requirements, and envisioning long-term value creation is challenging. Despite a strong inclination toward stakeholder focus, pressure from shareholders and directors to achieve short-term targets may sometimes overshadow long-term goals (Jones-Khosla & Gomes, 2023). To successfully navigate these challenges, organisations must proactively engage with a diverse set of stakeholders, including investors, regulators, and local communities, to foster trust and garner support for long-term ESG initiatives.

3.3 Discussion of Implications

This chapter highlights the potential of ESG as a strategic HR approach impacting talent acquisition, retention, employee engagement, and corporate culture. Nonetheless, three significant challenges must be tackled: cultural awareness and adaptation, limited understanding of the ESG framework, and balancing short-term and long-term objectives.

3.3.1 Implications for Academia in Asia

The adoption of ESG principles across Asia is shaped by distinct cultural contexts, yet existing research has not fully explored these nuances. Future studies should examine how cultural values influence ESG integration in HR strategies, providing comparative insights across different Asian countries. Understanding these variations can help organisations tailor their approaches to align with local expectations and business environments. Additionally, research should investigate gaps in ESG framework comprehension among businesses and HR professionals. Addressing these knowledge gaps through targeted educational initiatives can enhance ESG implementation and long-term sustainability.

Further empirical research is needed to evaluate the effectiveness of ESG-focused HR strategies on key workforce outcomes, including talent acquisition, retention, employee engagement, and corporate culture. While conceptual discussions highlight ESG's potential benefits, quantitative and qualitative studies can validate its long-term impact. Cross-industry and regional comparisons can offer deeper insights into sector-specific challenges and opportunities, enabling organisations to refine their ESG-driven HR strategies.

Leadership also plays a critical role in shaping ESG adoption. Prior studies (e.g., Yue et al., 2019) suggest that transformational leadership and transparent communication foster employee trust during organisational transitions. Future research could examine how leadership styles influence employee confidence and engagement in ESG-driven workplaces. Additionally, exploring effective communication strategies to embed ESG values within organisational culture could provide actionable insights for HR practitioners.

In the academic sphere, integrating ESG into business and HRM curricula is essential for preparing future professionals. Case studies of successful ESG adoption in Asian companies can provide valuable learning resources, while faculty training and development programs can equip educators with the expertise to teach ESG-centric HR strategies effectively. As ESG continues to evolve, collaborative research between academia and industry will be crucial in shaping HR practices that align with sustainability and responsible business leadership.

3.3.2 Implications for Practice in Asia

For businesses operating in Asia, integrating ESG principles into HR strategies is no longer optional; it is a necessity for long-term sustainability and competitiveness. The successful adoption of ESG-driven HRM requires organisations to align HR policies with sustainability objectives, ensuring they reflect both global ESG priorities and local cultural contexts. A crucial first step is embedding ESG into talent management by incorporating clear performance indicators such as workforce diversity, sustainability training participation, and employee well-being assessments. These metrics serve as measurable benchmarks that allow organisations to assess the impact of ESG initiatives on employee engagement, recruitment efficiency, and retention rates.

Despite its benefits, ESG implementation often faces resistance from employees who may perceive such initiatives as disruptive or misaligned with traditional business priorities. Overcoming this challenge requires structured change management, where organisations emphasise transparent internal communication, continuous employee education, and leadership advocacy (Jin & Kim, 2022). By integrating ESG into learning and development programs, HR professionals can ensure that employees at all levels understand the significance of sustainability in the workplace and are equipped with the necessary skills to support ESG objectives. Additionally, cultivating an ESG-centric organisational culture, where sustainability and social responsibility are embedded into core values, can foster stronger employee commitment and alignment with corporate goals.

Navigating the complex and evolving ESG regulatory landscape in Asia presents another challenge for businesses. With governments increasingly enforcing sustainability disclosure requirements, companies must remain informed about policy changes and proactively adapt their HR policies to ensure compliance. This necessitates closer collaboration between HR, legal, and compliance teams to integrate ESG governance into HR frameworks while maintaining alignment with industry best practices. Furthermore, multinational corporations operating across multiple Asian markets should consider establishing regional ESG governance frameworks to maintain consistency while addressing country-specific regulatory variations.

Technology plays an increasingly critical role in ESG implementation within HRM. AI and data analytics can enhance the tracking and evaluation of ESG-driven HR initiatives, from monitoring workforce sustainability to assessing diversity and inclusion efforts. However, as businesses leverage digital tools to streamline ESG integration, they must also prioritise ethical considerations. Ensuring transparency in AI-driven HR decisions, mitigating biases in recruitment algorithms, and upholding fair employment practices are essential for maintaining trust and accountability.

By embedding ESG into HR strategies, fostering employee engagement through education and culture-building, maintaining regulatory compliance, and leveraging technology responsibly, organisations in Asia can position themselves as leaders in sustainable workforce management. Through these efforts, companies can not only enhance their employer brand and operational resilience but also contribute to a more ethical, inclusive, and environmentally responsible future of work.

3.4 Future Directions

The future of work in Asia will be shaped by the increasing integration of ESG principles into corporate and HR strategies, driving transformations in workplace structures, workforce priorities, and leadership expectations. As businesses navigate this evolving landscape, they must anticipate and adapt to key changes that will redefine the employment experience in the coming years. One of the most significant shifts is the growing demand for purpose-driven work, particularly among Millennials and Gen Z employees. These younger generations increasingly seek employers whose values align with their own, particularly in areas such as sustainability, ethical governance, and social responsibility (Chatzopoulou & de Kiewiet, 2021; Tkalac Verčič & Verčič, 2024). Companies that effectively embed ESG into their organisational culture will have a competitive advantage in attracting and retaining top talent, as employees become more selective in choosing workplaces that reflect their personal and professional aspirations.

In parallel, the nature of work itself is undergoing profound transformation, driven by digitalisation and the expansion of hybrid and remote work models. ESG considerations will play a crucial role in shaping these changes, influencing how organisations design workspaces, manage employee well-being, and promote work-life balance. Future workplaces will be more flexible and digitally integrated, with companies investing in sustainable infrastructure, green building practices, and energy-efficient office environments (Asif et al., 2024). At the same time, hybrid work arrangements will require businesses to establish policies that align with ESG commitments, ensuring that remote work strategies contribute to lower carbon footprints while maintaining employee engagement and inclusivity (Shukor & Ghazali, 2025).

Regulatory developments will also continue to drive the evolution of work in Asia, with governments enforcing stricter ESG disclosure and compliance requirements. Organisations will need to integrate sustainability governance into workforce planning, ensuring that HR practices align with emerging regulatory frameworks. This shift will likely lead to greater standardisation in ESG reporting, increased corporate transparency, and a stronger focus on measuring the social and environmental impact of employment policies (Singhania & Saini, 2022). As a result, HR leaders must take on a more strategic role in sustainability reporting, ensuring that employment practices are not only legally compliant but also aligned with global ESG expectations.

Leadership in the future of work will also require a fundamental shift in mindset and competencies. ESG-conscious leadership will become a defining trait of successful organisations, as businesses seek executives who can balance profitability with long-term social and environmental impact (Ferrarini, 2021). Ethical decision-making, stakeholder engagement, and sustainability-driven leadership development will become critical competencies for managers across industries. The role of Chief Sustainability Officers (CSOs) will expand, with HR playing a key role in integrating sustainability expertise into executive leadership pipelines. Companies that invest in leadership training programs focused on ESG principles will be better positioned to navigate the complexities of the evolving business landscape (Foster et al., 2023).

Finally, technological advancements will further shape the future of work, with AI and automation playing a pivotal role in ESG-driven HR strategies. AI-powered HR analytics will enable predictive workforce planning, ESG risk assessments, and real-time employee sentiment analysis regarding sustainability efforts (Chawla et al., 2025). Automation will support ESG performance tracking, recruitment fairness, and sustainability-driven workforce planning. However, businesses must ensure that these technologies are implemented ethically, with safeguards in place to prevent algorithmic bias, enhance transparency, and uphold responsible employment practices (Tripathi & Kumar, 2025).

As organisations prepare for the future of work in Asia, those that proactively embrace ESG principles will gain a strategic advantage in building a resilient, engaged, and sustainability-driven workforce. The integration of ESG into HRM is not merely a compliance requirement; it is a transformative force that will define the long-term success of businesses across the region. By aligning workforce strategies with sustainability objectives, fostering inclusive and purpose-driven workplace cultures, and leveraging technology responsibly, companies can position themselves as leaders in shaping the future of work in an ESG-centric world.

3.5 Conclusion

The integration of ESG principles into HR strategies is no longer a peripheral initiative but a fundamental driver of organisational success in Asia's evolving workforce landscape. ESG-driven HRM enhances talent acquisition, employee retention, engagement, and corporate culture, positioning organisations to thrive in an era where sustainability, ethics, and social responsibility are core business imperatives. By embedding ESG into HR policies, organisations can create a purpose-driven work environment that aligns with both employee values and emerging regulatory requirements.

At the same time, businesses must navigate challenges such as cultural adaptation, evolving compliance frameworks, and resistance to change. A structured approach to change management, supported by leadership commitment and clear ESG performance metrics, is essential for translating sustainability ambitions into effective workforce strategies. Furthermore, leveraging AI and data analytics can help HR professionals track ESG performance, optimise workforce planning, and enhance diversity and inclusion efforts, provided that ethical considerations remain central to technological adoption.

Looking ahead, ESG principles will continue to reshape the future of work in Asia, influencing leadership expectations, employment structures, and regulatory landscapes. Companies that proactively adapt to digitalisation, hybrid work trends, and sustainability-driven governance models will gain a competitive edge in attracting and retaining talent. Moreover, ESG-conscious leadership will become an essential competency, requiring executives to balance financial performance with long-term social and environmental impact.

Ultimately, businesses that embed ESG into HRM will not only strengthen their employer brand but also contribute to a more responsible and resilient future workforce. The evolving work environment in Asia demands that organisations move beyond compliance-driven approaches and embrace ESG as a strategic enabler of sustainable workforce development. By fostering inclusive workplace cultures, investing in ESG leadership development, and integrating ethical AI-driven workforce strategies, companies can position themselves as leaders in the next era of work.

Highlights

- ESG-driven HR strategies are shaping the future of work in Asia, influencing talent management, leadership, and workplace structures.

- Digitalisation, AI, and hybrid work models are transforming HR practices.
- Evolving ESG regulations and leadership expectations will redefine workforce strategies.
- Companies embedding ESG into HRM will gain a long-term competitive edge.

References

Asif, M., Naeem, G., & Khalid, M. (2024). Digitalization for sustainable buildings: Technologies, applications, potential, and challenges. *Journal of Cleaner Production*, 141814. https://doi.org/10.1016/j.jclepro.2024.141814

Asif, M., Searcy, C., & Castka, P. (2023). ESG and Industry 5.0: The role of technologies in enhancing ESG disclosure. *Technological Forecasting and Social Change, 195*, 122806. https://doi.org/10.1016/j.techfore.2023.122806

Atkins, J., Doni, F., Gasperini, A., Artuso, S., La Torre, I., & Sorrentino, L. (2023). Exploring the effectiveness of sustainability measurement: Which ESG metrics will survive COVID-19? *Journal of Business Ethics, 185*(3), 629–646. https://doi.org/10.1007/s10551-022-05183-1

Bukreeva, A., & Grishunin, S. (2023). Estimation of impact of ESG practices' performance and their disclosure on company's value. *Procedia Computer Science, 221*, 322–329. https://doi.org/10.1016/j.procs.2023.07.044

Chatzopoulou, E., & de Kiewiet, A. (2021). Millennials' evaluation of corporate social responsibility: The wants and needs of the largest and most ethical generation. *Journal of Consumer Behaviour, 20*(3), 521–534. https://doi.org/10.1002/cb.1882

Chawla, M., Sharma, P., Gautam, S., & Popescu, C. R. G. (2025). Sustainability in human capital management: Shaping the future with technology. In *Sustainable management practices for employee retention and recruitment* (pp. 265–290). IGI Global Scientific Publishing

Cho, Y., Kim, S., You, J., Moon, H., & Sung, H. (2021). Application of ESG measures for gender diversity and equality at the organizational level in a Korean context. *European Journal of Training and Development, 45*(4/5), 346–365. https://doi.org/10.1108/EJTD-05-2020-0090

Edelman. (2022). *Edelman trust barometer 2022—Global report.* Retrieved January 3, 2024, from https://www.edelman.com/sites/g/files/aatuss191/files/2022-01/2022%20Edelman%20Trust%20Barometer%20FINAL_Jan25.pdf.

Ferrarini, G. (2021). Redefining corporate purpose: Sustainability as a game changer. In D. Busch, G. Ferrarini, & S. Grünewald (Eds.), *Sustainable finance in Europe. EBI studies in banking and capital markets law.* Palgrave Macmillan. https://doi.org/10.1007/978-3-030-71834-3_4

Foster, B. P., Manikas, A. S., & Kroes, J. R. (2023). Which diversity measures best capture public company value? *Corporate Social Responsibility and Environmental Management, 30*(1), 236–247. https://doi.org/10.1002/csr.2351

Jakob, E. A., Steinmetz, H., Wehner, M. C., Engelhardt, C., & Kabst, R. (2022). Like it or not: When corporate social responsibility does not attract potential applicants. *Journal of Business Ethics, 178*(1), 105–127. https://doi.org/10.1007/s10551-021-04960-8

Jasni, N. S., Yusoff, H., Zain, M. M., Md Yusoff, N., & Shaffee, N. S. (2020). Business strategy for environmental social governance practices: Evidence from telecommunication companies in Malaysia. *Social Responsibility Journal, 16*(2), 271–289. https://doi.org/10.1108/SRJ-03-2017-0047

Jin, M., & Kim, B. (2022). The effects of ESG activity recognition of corporate employees on job performance: The case of South Korea. *Journal of Risk and Financial Management, 15*(7), 316. https://doi.org/10.3390/jrfm15070316

Johnson, R. D., Stone, D. L., & Lukaszewski, K. M. (2021). The benefits of eHRM and AI for talent acquisition. *Journal of Tourism Futures, 7*(1), 40–52. https://doi.org/10.1108/JTF-02-2020-0013

Jones-Khosla, L. A., & Gomes, J. F. S. (2023). Purpose: From theory to practice. *Global Business and Organizational Excellence, 43*(1), 90–103. https://doi.org/10.1002/joe.22203

Karwowski, M., & Raulinajtys-Grzybek, M. (2021). The application of corporate social responsibility (CSR) actions for mitigation of environmental, social, corporate governance (ESG) and reputational risk in integrated reports. *Corporate Social Responsibility and Environmental Management, 28*(4), 1270–1284. https://doi.org/10.1002/csr.2137

Kim, R. C. (2022). Rethinking corporate social responsibility under contemporary capitalism: Five ways to reinvent CSR. *Business Ethics, the Environment & Responsibility, 31*(2), 346–362. https://doi.org/10.1111/beer.12414

Kotsantonis, S., & Serafeim, G. (2019). Four things no one will tell you about ESG data. *Journal of Applied Corporate Finance, 31*(2), 50–58. https://doi.org/10.1111/jacf.12346

Liew, S. (2023, March 27). *Embracing sustainable business practices gains momentum in Asia.* HRM Asia. Retrieved January 3, 2024, from https://hrmasia.com/embracing-sustainable-business-practices-gains-momentum-in-asia/

Ling, Y. H. (2019). Cultural and contextual influences on corporate social responsibility: A comparative study in three Asian countries. *Cross Cultural & Strategic Management, 26*(2), 290–310. https://doi.org/10.1108/CCSM-02-2018-0024

Malhotra, Y., & Pachauri, V. (2023). Employee engagement in ESG practices: A way to sustainability. In S. Singh, A. Kumari, & P. Haldar (Eds.), *Digital disruption and environmental, social & governance* (pp. 34–46). Book Bazooka Publication.

Mihardjo, L. W., Jermsittiparsert, K., Ahmed, U., Chankoson, T., & Iqbal Hussain, H. (2020). Impact of key HR practices (human capital, training and rewards) on service recovery performance with mediating role of employee commitment of the Takaful industry of the Southeast Asian region. *Education+ Training, 63*(1), 1–21. https://doi.org/10.1108/ET-08-2019-0188

Mohammad, W. M. W., & Wasiuzzaman, S. (2021). Environmental, Social and Governance (ESG) disclosure, competitive advantage and performance of firms in Malaysia. *Cleaner Environmental Systems, 2*, Article 100015. https://doi.org/10.1016/j.cesys.2021.100015

Mojodomo. (2022). Asia's workforce awakens to ESG. Retrieved April 5, 2024, from https://uploads-ssl.webflow.com/6331610ee5a254c8811d754b/63480b48bbae5ddc217b92bc_Asias%2BWorkforce%2BAwakens%2Bto%2BESG.pdf

Nekhili, M., Boukadhaba, A., & Nagati, H. (2021). The ESG–financial performance relationship: Does the type of employee board representation matter? *Corporate Governance: An International Review, 29*(2), 134–161. https://doi.org/10.1111/corg.12345

Ng, T. H., Lye, C. T., Chan, K. H., Lim, Y. Z., & Lim, Y. S. (2020). Sustainability in Asia: The roles of financial development in environmental, social and governance (ESG) performance. *Social Indicators Research, 150*, 17–44. https://doi.org/10.1007/s11205-020-02288-w

Parameswar, N., Hasan, Z., Shri, C., & Saini, N. (2024). Exploring the barriers to ESG adoption using modified TISM approach. *Kybernetes, 53*(12), 5775–5800. https://doi.org/10.1108/K-05-2023-0888

Piao, X., Xie, J., & Managi, S. (2022). Environmental, social, and corporate governance activities with employee psychological well-being improvement. *BMC Public Health, 22*, 1–12. https://doi.org/10.1186/s12889-021-12350-y

Poppelwell, E., & Overton, J. (2022). Culture, scale and the adaptation of good governance: Insights from SOEs in Samoa and Tonga. *World Development Perspectives, 28*, Article 100468. https://doi.org/10.1016/j.wdp.2022.100468

PwC. (2020). *Environmental, social and governance (ESG) in Asia—Asset and wealth management.* PricewaterhouseCoopers. Retrieved January 3, 2024, from https://www.pwc.com/sg/en/asset-management/assets/environmental-social-and-governance-in-asia-awm.pdf

PwC. (2023). *Sustainability counts II—State of sustainability reporting in Asia Pacific.* PricewaterhouseCoopers. Retrieved April 10, 2024, from https://www.pwc.com/sg/en/publications/assets/page/sustainability-counts-2023.pdf

Saks, A. M. (2019). Antecedents and consequences of employee engagement revisited. *Journal of Organizational Effectiveness: People and Performance, 6*(1), 19–38. https://doi.org/10.1108/JOEPP-06-2018-0034

Saluto, V. (2024). *Navigating the future: A deep dive into Asia's ESG landscape and dynamics.* Asia Business Outlook. Retrieved April 5, 2024, from https://www.asiabusinessoutlook.com/industry-experts/navigating-the-future-a-deep-dive-into-asia-s-esg-landscape-and-dynamics-nwid-5851.html

Schoenmaker, D., & Schramade, W. (2019). Investing for long-term value creation. *Journal of Sustainable Finance & Investment, 9*(4), 356–377. https://doi.org/10.1080/20430795.2019.1625012

Shrotryia, V. K., & Dhanda, U. (2019). Content validity of assessment instrument for employee engagement. *SAGE Open, 9*(1), 2158244018821751. https://doi.org/10.1177/21582440188217

Shukor, S. F. A., & Ghazali, F. (2025). Embracing ESG in the era of remote work: legal requirement and benefits. In *Implementing ESG frameworks through capacity building and skill development* (pp. 365–380). IGI Global Scientific Publishing.

Singhania, M., & Saini, N. (2022). Quantification of ESG regulations: A cross-country benchmarking analysis. *Vision, 26*(2), 163–171. https://doi.org/10.1177/0972262921105417

Stahl, G. K., Brewster, C. J., Collings, D. G., & Hajro, A. (2020). Enhancing the role of human resource management in corporate sustainability and social responsibility: A multi-stakeholder, multidimensional approach to HRM. *Human Resource Management Review, 30*(3), Article 100708. https://doi.org/10.1016/j.hrmr.2019.100708

Tan, P. (2022, January 11). *ESG: A lever to attract and retain talent in Singapore.* JLL. Retrieved April 5, 2024, from https://www.jll.com.au/en/trends-and-insights/research/esg-a-lever-to-attract-and-retain-talent-in-singapore

Tan, J. (2023, February 21). *Foster a sustainable future: Reflecting ESG values in the workplace.* HRM Asia. Retrieved January 3, 2024, from https://hrmasia.com/fostering-a-sustainable-future-reflecting-esg-values-in-the-workplace/

Tavanti, M. (2023). EESG organizational sustainability management. In *Developing sustainability in organizations. Sustainable development goals series* (pp. 171–218). Palgrave Macmillan. https://doi.org/10.1007/978-3-031-36907-0_7

Tkalac Verčič, A., & Verčič, D. (2024). Attitudes toward sustainable development and employer brands: Comparing generations X, Y and Z in two countries. *Corporate Communications: An International Journal, 30*(2), 355–371. https://doi.org/10.1108/CCIJ-11-2023-0161

Torres, L., Ripa, D., Jain, A., Herrero, J., & Leka, S. (2023). The potential of responsible business to promote sustainable work–An analysis of CSR/ESG instruments. *Safety Science, 164*, Article 106151. https://doi.org/10.1016/j.ssci.2023.106151

Tripathi, A., & Kumar, V. (2025). Ethical practices of artificial intelligence: a management framework for responsible AI deployment in businesses. *AI and Ethics*, 1–12. https://doi.org/10.1007/s43681-025-00670-3

Tumasjan, A., Kunze, F., Bruch, H., & Welpe, I. M. (2020). Linking employer branding orientation and firm performance: Testing a dual mediation route of recruitment efficiency and positive affective climate. *Human Resource Management, 59*(1), 83–99. https://doi.org/10.1002/hrm.21980

Tunio, R. A., Jamali, R. H., Mirani, A. A., Das, G., Laghari, M. A., & Xiao, J. (2021). The relationship between corporate social responsibility disclosures and financial performance: A mediating role of employee productivity. *Environmental Science and Pollution Research, 28*, 10661–10677. https://doi.org/10.1007/s11356-020-11247-4

United Nations Global Compact. (2021). Principles for responsible investment. Retrieved January 3, 2024, from https://www.unpri.org/download?ac=10948

Yue, C. A., Men, L. R., & Ferguson, M. A. (2019). Bridging transformational leadership, transparent communication, and employee openness to change: The mediating role of trust. *Public Relations Review, 45*(3), Article 101779. https://doi.org/10.1016/j.pubrev.2019.04.012

Zahari, A. I., Said, J., Muhamad, N., & Ramly, S. M. (2024). Ethical culture and leadership for sustainability and governance in public sector organisations within the ESG framework. *Journal of Open Innovation: Technology, Market, and Complexity, 10*(1), Article 100219. https://doi.org/10.1016/j.joitmc.2024.100219

Zainee, I. A., & Puteh, F. (2020). Corporate social responsibility impact on talent retention among Generation Y. *Revista De Gestão, 27*(4), 369–392. https://doi.org/10.1108/REGE-06-2019-0070

Jing Yi Yong is a Senior Lecturer of the School of Management and Marketing at Taylor's University. She has received her Doctorate degree from Universiti Sains Malaysia. She has been published in various book chapters and scholarly journals such as Journal of Cleaner Production, Business Strategy and Environment, Benchmarking: An International Journal, Industrial and Commercial Training, Journal of Management Development, and International Journal of Engineering and Technology. She also serves as a reviewer of international journals and international conferences. Her research interests include green human resource management, intellectual capital, ESG and sustainability.

Woon Leong Lin is an Associate Professor at the Faculty of Business and Law, Taylor's University. Since completing his Ph.D. from Universiti Putra Malaysia, and his MBA from East Anglia University, UK, Lin's research focus has been in the areas of corporate political strategy, business ethics, and corporate social responsibility. Within ethics, he explores factors influencing business and organizational ethics, and ethical behaviour. His other key research area in corporate social responsibility focuses on green environmental practices and its relationship to organizational practices and performance. He has published numerous articles in refereed journals, including journals indexed by Thomson Web of Science (SSCI) and Scopus. His publications include Business Strategy and Environment, Corporate Social Responsibility and Environmental Management, The North American Economics and Finance, Sustainability, International Journal of Economics and Management, International Financial Study, Industrial Marketing Management, Administrative Sciences, Journal of Cleaner Production, Geojournal, and Environmental Science and Pollution Research.

Thanuja Rathakrishnan is a Senior Lecturer at Taylor's Business School, Taylor's University, Malaysia. She earned her Ph.D. in Business Economics at Universiti Putra Malaysia (UPM). Prior to this, she obtained a Master of Business Administration from Putra Business School, UPM, Malaysia and was conferred a bachelor's degree in Communication and Management by Upper IOWA University, USA. Thanuja also partakes in research and consultancy projects with government and non-government organisations. Her areas of interest include, but not limited to entrepreneurship, small businesses, and family businesses. In addition to her academic role, Thanuja is also a Stream Coordinator in her organisation, an Editorial Review Board Member of the Journal of Responsible Tourism Management, and a Local Virtual Volunteer for AIESEC.

Thivashini B. Jaya Kumar is a Senior Lecturer at the School of Management & Marketing, Taylor's Business School, Taylor's University, Malaysia. She was awarded Bachelor of Business from Victoria University, Australia and Master in Management from Taylor's University. Prior to entering academia, she was a business analyst with 3 + years of experience. She holds a Ph.D. in Business (Marketing) with a research interest in service marketing, consumer behaviour, international business, and entrepreneurship.

Feranita Feranita is an Associate Professor in Entrepreneurship and Innovation at Nottingham University Business School China, University of Nottingham Ningbo China. She received her Ph.D. from Lancaster University, UK. Prior to that, she obtained a Master of Research in Management Sciences from ESADE Business School, Spain, and an MBA from Nanyang Business School, Singapore. Primarily focusing on qualitative research method, her research centres around family business, entrepreneurship, innovation, and university-industry collaborations. She has published in journals such as British Journal of Management, R&D Management, and Journal of Family Business Strategy. Her publications also include family business cases in SAGE Business Cases, book, book chapters in entrepreneurship, social entrepreneurship and sustainability, as well as media articles. As a certified professional coach, she coaches startups and business families.

Nelvin XeChung Leow is a distinguished faculty member at Taylor's University in Malaysia. His expertise lies in the field of environmental management, with a primary focus on Travel Behaviour Modelling, Smart Mobility, and Sustainability. He is not only a dedicated educator but also actively contributes to the academic community as a reviewer for several esteemed journals. His research and teaching endeavours play a pivotal role in advancing our understanding of sustainable practices and smart mobility solutions, making him a valuable asset in the pursuit of a more environmentally conscious future.

Part II
Future Work Dynamics in Asia

Chapter 4
Navigating the Future of Work: Managing a Multigenerational Workforce

Clusters of Future Studies: Corporate Foresight; Humanity at the Limen

Sheena D. C. Doria-De Vera

Abstract The world today, Asia included, is experiencing a demographic shift. Complex workforce demographics calls for new human resource strategies to respond to the needs of a wide range of generations and backgrounds in organisations. For the first time in history, five generations are active in the workplace: Silent Generation, Baby Boomers, Generation X, Millennials, and Generation Z. Aside from the preponderance of different generations of workers, the United States Census Bureau estimates that in 2060, 42 over 54 countries in Asia will have 14% share of 65s to the total population. Most of the countries in Asia will proceed to age more rapidly. Thus, this conceptual paper delves into the dynamic landscape of the future of work in Asia focusing on managing a multigenerational workforce. This paper used typology as the key methodology to describe the unique characteristics, work styles, preferred communication styles, preferred leadership styles, and preferred rewards of each generation. These differences between the generations emphasise the need for tailored management approaches to foster harmonious relationships and productive workplaces in Asia. This paper also intends to provide recommendations for organisations aiming to thrive in the Asian market by managing the generational diversity effectively.

Keywords Generational diversity · Multigenerational workforce · Generation Z · Generation Alpha · Ageing workforce · Workforce trends in Asia

S. D. C. Doria-De Vera (✉)
Institute of Management, College of Social Sciences, University of the Philippines, Baguio City, Philippines
e-mail: sddoria@up.edu.ph

C. Y. Fung et al. (eds.), *Future of Work in Asia*,
https://doi.org/10.1007/978-3-031-95804-5_4

4.1 Introduction

The world today, Asia included, is experiencing a demographic shift. The world's population is expected to reach the highest point at 9.62 billion in 2050, a 1.5 billion increase from 8.06 billion in 2023, according to the World Bank (n.d.). With this population, the biggest chunk of about 59.27% come from Asia (United Nations, 2024).

In some regions of the world, young people aged 15 to 24 years comprise a large and rapidly growing segment of the population. Many of the countries in the Asian region witnessed an increase in a similar bulge of young people followed by a bulge of working-age people (25 to 64 years). Fertility rates continue to decline, and older persons (65 years and above) today comprise the world's fastest-growing age group. Globally, for the first time in 2018, older persons outnumbered children under the age of five, and by 2050, older persons will outnumber adolescents and youth (UN, n.d.). These claims are supported by actual population data of World Bank Data where Asia is a melting pot of young societies, aging societies, aged societies, and one super-aged society.

One way to group age cohorts is through Generations. United Nations (2001) define generation as groups of persons who are at the same age at a particular time and who have passed or are passing through life, experiencing changes at the same time. For the first time in history, five generations are active in the workplace: Silent Generation, Baby Boomers, Generation X, Millennials and Generation Z The younger generation, Millennials and Generation Z, are expected to make up 50% of the Asia–Pacific region by 2025 (Kim et al., 2020). The preponderance of different generations of workers and the ageing workforce poses both unprecedented opportunities and challenges for employers and organisations. Differences between the older and younger generations could potentially create chaos in the workplace (Anderson et al., 2017). Managers nowadays have a challenging role in catering different needs of its employees for them to feel nurtured and remain committed to the organisation. After all, humans, even belonging to the same generations, are multi-layered and complex; therefore, the goal is to create a nurturing environment where this complexity can flourish (Singh, 2023).

Thus, this chapter delves into the dynamic landscape of the future of work in Asia focusing on managing a multigenerational workforce. This examines the different generations' work values or work styles and their preferred communication styles, leadership styles, and rewards, emphasising the need for tailored management approaches to foster harmonious relationships and productive workplaces in the diverse cultural context of Asia. Special attention is given to the distinctive traits of Millennials and Generation Z that will make-up half of the workforce in the near future. This paper also intends to provide recommendations for organisations aiming to thrive in the Asian market by managing the generational diversity effectively.

4.2 Methodology

This chapter used the typology approach as the primary methodology in examining the characteristics of the multigenerational workforce in Asia. Typology is one of the types of research design in conceptual papers discussed by Jaakkola (2020). Typology involves creating a classification system based on shared characteristics, in this case, different generations' work values or work styles and their preferred communication styles, leadership styles, and rewards. This approach would help in the recognition of intergenerational patterns, which would provide a more detailed understanding of how to manage a diverse workforce in Asia. In addition, to discuss the issues related to an ageing workforce, secondary data was incorporated from sources like World Bank, United Nations, U.S. Census Bureau, and other journals. The demographic profiles of the Asian countries were summarised and analysed using descriptive statistics.

4.3 Key Concepts and Issues

This part discusses the different key concepts and issues about managing the multigenerational workforce in Asia. It starts with the discussion of the concept of generational theory. This will be followed by the discussion and summary of characteristics of four generations present in the workplace: Baby Boomers, Generation X, Millennials and Generation Z. The last section focuses on the demographic profile of Asian workforce by classifying countries as young, aging, aged, or superaged societies.

4.3.1 The Generations

The concept of generational theory was popularised by Neil Howe and William Strauss in the 1990s in their book "Generations: The History of America's Future, 1584 to 2069". They emphasised that generations affect society. The theory describes a recurring cycle of same-aged groups with specific behavioural patterns that change every 20 years. According to this theory, every 80-years, each of the four generations is associated with a crisis that impacts the current social order and then creates a new one (Flavia, 2019). For that matter, this theory pertains to the baby boomers, Generation X, Millennials and Generation Z.

However, McCrindle and Wolfinger (2020) argued that this definition is already irrelevant because cohorts are changing so quickly in response to new technologies, changing career and study options, and shifting societal values. Thus, twenty years is far too broad to define a generation. Up to this date, there is no consensus on the range of birth years for each generation. For simplification, the author will use the one from Pew Research Center and US Census Bureau as cited by Beresford Research (2024).

Table 4.1 The multigenerational workforce (Pew Research Center and US Census Bureau as cited by Beresford Research, 2024 and author's calculations)

Generation	Birth years	Age as of 2024
Silent Generation	1928–1945	79–96
Baby Boomers	1946–1964	60–78
Generation X	1965–1980	44–59
Millennials	1981–1996	28–43
Generation Z	1997–2012	12–27

Table 4.1 shows the different generations' birth years, some of their common traits and their age as of 2024.

The oldest generation and are almost out of the workforce are the Silent Generation. Some countries still hire aged 70 + and most probably hire more (such as Japan). This is because of the ageing population, but most literature no longer includes them since the youngest age is 79 years old. Thus, the next part of this chapter will describe the different work preferences of the four major generations: Baby Boomers, Generation X, Generation Y and Generation Z.

4.3.1.1 Baby Boomers

This cohort is born from 1946 to 1964 and as of 2024, they are between 60 and 78 years old. They are known for valuing loyalty and commitment, dedication to their work, relationship building with colleagues, and taking a process-oriented perspective in how they view work (Rathi & Kumar, 2023). They are characterised as dependable and driven individuals, prioritising liberty and determination in their careers. However, the strong work ethic of the baby boomers can create challenges in achieving work-life balance as they fear absenteeism may put their job or position at risk. They are convinced that success is earned through hard work and support the idea of seniority and lifetime employment. The workplace values of baby boomers and younger generations often clash due to differences in work ethics and working hours (Ekambaram, 2014; Gibson et al., 2009 quoted in Rathi & Kumar, 2023; Coulter & Faulkner, 2014 quoted in Rathi & Kumar, 2023). In terms of communication, baby boomers strongly prefer one-on-one interactions and favour face-to-face communication over modern methods (Rathi & Kumar, 2023).

In terms of leadership styles, baby boomers tend to lean towards task-oriented leaders who value cooperation and a harmonious work environment. They also want leaders who recognize and appreciate their hard work, enthusiasm, and contributions to the organisational initiatives (Lockwood, 2004 as cited in Rathi & Kumar, 2023). Although receptive to participative leadership approaches and appreciate being involved in decision-making processes, baby boomers show their respect for authority within organisations (Bertsch et al., 2022). In terms of reward preferences, baby boomers prioritise caregiving and retirement benefits, and comprehensive healthcare plans. Moreover, they also value ongoing training opportunities and workplace flexibility (California Employers Association, 2023; Obsidian HR, 2024).

4.3.1.2 Generation X

This cohort were born between 1965 to 1980 and as of 2024 they are between 44 to 59 years old. They are known for their distinct work values and ethics. They are characterised as fun-loving, informal, independent, and cyclical individuals who strive to achieve a balance between work and family (Gibson et al., 2009 as cited in Rathi & Kumar, 2023). This generation is capable of using technology, values work-life balance and tends to be more sceptical and distrust authority figures (Coulter & Faulkner, 2014 as cited in Rathi & Kumar, 2023). In terms of communication, Generation X prefers email for exchanging information over one-on-one interaction. However, utilising face-to-face meetings and phone calls are still relevant to them (Rathi & Kumar, 2023). Leadership-wise, this generation favours leaders who encourage their participation in decision-making processes, value honesty, prioritise work-life balance, and invest in training opportunities (Rathi & Kumar, 2023). The independent trait of Generation X makes them see the value of autonomy and flexibility and less reliant on leaders. However, they still appreciate fair, straightforward, and competent leadership (Bertsch et al., 2022).

When it comes to reward, Generation X recognizes the importance of quick recognition for their efforts. This can take various forms, including pay, praise, promotion, and title (Rathi & Kumar, 2023). It can also come along with remote work options, and flexible scheduling, and benefits like caregiving assistance, retirement planning, and other work-life balance initiatives (California Employers Association, 2023; Obsidian HR, 2024).

4.3.1.3 Millennials

Also known as Generation Y, this is the generation who were born between 1981 and 1996 and are the most dominant generation in the workplace now. Millennials are currently between 28 and 43 years of age. They were once known as the most misunderstood generation demonstrating distinct work values and ethics to the workplace. They are renowned for their tech-savviness and confidence, a strong drive for financial success, and for their propensity to change jobs (Adkins, 2024; Coulter & Faulkner, 2014 as cited in Rathi & Kumar, 2023).

With regards to communication, millennials deem it most effective when it is consistent, clear, and delivered through email. They value broad-minded and regular engagement with their leaders (Rathi & Kumar, 2023). Millennials are characterised as technologically fluent individuals, often relying on digital information tools and value collaborative communication methods (Eisner, 2005 as cited in Bertsch et al., 2022). The preferred leaders of millennials are those who set clear expectations, long-term goals, and provide opportunities for growth and development through accessible training programs (Bertsch et al., 2022). They are looking for strong leadership and appreciate leaders who utilise technology for regular updates on job-related matters and cultivate a positive work environment (Martin, 2005 as cited in Bertsch et al., 2022).

In terms of rewards, millennials give precedence to maintaining work-life balance, including flexible scheduling and remote work opportunities. Also important are financial aspects like financial guidance and loan repayment assistance. They also appreciate parental leaves, ongoing training, and other opportunities for career advancement (California Employers Association, 2023; Obsidian HR, 2024). Being portrayed as softer and caring, millennials value workplace fun, positive feedback, while having intolerance to dishonesty. They value genuine care and honesty from their leaders and cherish being recognized publicly for their achievements (Bertsch et al., 2022).

4.3.1.4 Generation Z

This generation was born between 1997 and 2012 and is currently the youngest generation in the workplace. They are currently between 12 and 27 years old as of 2024. This generation demonstrates unique attributes that differentiate them from previous generations. They are known for their strong work ethic, multitasking ability, optimism, and proficiency in using new technologies (Alexander, 2018 as cited in Rathi & Kumar, 2023; Dolot, 2018 as cited in Rathi & Kumar, 2023). They recognize the value of work-life balance, independence, favour to learn on their own, and maintain boundaries between work and personal life (Rohei, 2019 as cited in Rathi & Kumar, 2023).

When it comes to communication, Generation Z or Gen Z strongly favours it being clear and concise. They value one-on-one interaction, both face-to-face and through online platforms, for the efficient and quick delivery of information (Rathi & Kumar, 2023). Leadership wise, Generation Z thrives in a workplace that provides growth opportunities and flexibility. They are grateful for mentors who can provide feedback and hands-on experiences that may help them further develop their skills (Walters, 2020).

More than just paycheck or traditional benefits like health insurance and retirement plans, Gen Z prefers rewards that may provide mental health support and diversity and inclusion initiatives (Obsidian HR, 2024). Also important for Gen Zs are remote work options, flexible work arrangements, and comprehensive employee assistance programs (California Employers Association, 2023). They are also considered the most progressive generation and see values on employers who support mental health and social justice issues actively beyond just offering benefits (Obsidian HR, 2024).

Table 4.2 summarises each generation's work styles, preferred communication styles, preferred leadership styles, and preferred rewards.

Table 4.2 Comparison of work preferences across generations (Adkins, 2023; Bertsch et al., 2022; California Employers Association, 2023; Obsidian HR, 2024; Rathi & Kumar, 2023)

	Baby Boomers	Generation X	Millennials	Generation Z
Work styles/work values/work ethics	Workaholics, "live to work", focus on work/career oriented, loyalty and commitment and relationship building, cutthroat and reliable, busy bees, career-oriented who give importance to liberty and determination and experiencing poor work-life balance, loyal to employer	"Work to live"; fun-loving, cynical, informal, and independent and look for more of a balance amid work and family, does not have faith in authority	Live first, then work; most likely generation to switch jobs/job hoppers, confident, optimistic, socially sensitive, technologically adept, ambitious, curious, and easily get bored and tired; softer, more caring generation,	Multitaskers, and efficient in utilising new technology as compared to other generations; not afraid to express their feelings, self-directed and prefers an independent approach to learning, gives value to work-life balance/work life integration; does not want to think about work after working hours
Preferred communication styles	One on one interaction, focus only on face-to- face communication: personal touch, handwritten notes	Electronic-mail communication, face-to-face, email, or call communication	Instant messaging applications, texts and emails; use technology communications to keep everyone informed about their jobs on a regular everyday basis	Prefers communication via technology, online communication more than in person, amiable communicator, texts/tweets with emoji for a shortcut; social media
Preferred leadership styles	Harmonious and team-oriented leadership styles, participative style of leadership, and enjoy getting involved in the decision-making process, high respect for authority, traditional hierarchy-based leadership styles	Participative decision-making style extreme need for autonomy and flexibility in their own lifestyles and jobs, this means that there is less need for leadership prefer a leadership style that is fair, straightforward, and competent	Collaborative and supportive leadership styles; like leaders who keep the workplace fun and like to be let known that they are doing excellent work, like to hear positive feedback and do not like to be lied to; likes being guided in their early state of their career	Transformational and authentic leadership; prefer broad-minded and steady interaction with their leaders; likes leaders who could connect via relations, communicate rationally, manage practically, and lead with direction and strategy

(continued)

Table 4.2 (continued)

	Baby Boomers	Generation X	Millennials	Generation Z
Reward preferences	Caregiving benefits, comprehensive health care plans, Retirement and healthcare benefits—including medical, dental, and vision care, other types of wellness benefits such as gym memberships or discounts on health services; tangible symbols of loyalty, commitment and service such as plaques and certificates	Caregiving benefits, comprehensive health care plans, Retirement and healthcare benefits—including medical, dental, and vision care Other types of wellness benefits such as gym memberships or discounts on health services' personal attention, promotion and recognition, work life-balance	Expects quick recognition via pay, praise, promotion, and title; flexible scheduling, more vacation time, remote work opportunities, ongoing training opportunities, career advancement opportunities; work-life integration, certifications to add to their resumes; upgraded resources	Expects quick recognition via pay, praise, promotion, and title; flexible scheduling, more vacation time, remote work opportunities, ongoing training opportunities, career advancement opportunities; work-life integration; mental health benefits; benefits that support diversity, equity, and inclusion

4.3.2 The Demographic Profile of Asian Workforce

There are a lot of significant demographic trends in the world in recent years but undeniably, increasing attention has been given to the ageing population of Asia. It has the highest population in the world and its population is also ageing faster than any region (United Nations DESA, 2019). Population ageing is an inevitable outcome of the demographic transition—the historical shift from higher to lower levels of fertility and mortality that yields a period of rapid population increase and, eventually, an older population that is much larger as a share of the total (United Nations DESA, 2023). By 2060, older Asians are projected to exceed 1.2 billion, which implies that one out of every ten people in the world will be an older Asian (U.S. Census Bureau, 2021).

This also implies a shrinking workforce. The population aged between 16 and 64 is expected to peak by 2045 and then will begin declining after that. Over the last three decades, the labour force participation rates of individuals in the region aged between 60 and 64 remained relatively high at more than 50%. Developed countries in Asia only reached this level recently. Older people in Asia are usually employed in low-wage farm and retail jobs. For example, 65.2% of people aged 60–64 in Vietnam and 58.9% in Bangladesh are working in the agriculture sector. Following subsistence agriculture, which means that 'nearly all of the crops or livestock they raised are used to maintain the farmer and the farmer's family, leaving little, if any, surplus for sale or trade' (Brittanica, n.d.), many older adults who continue to work are in the wholesale and retail sector, especially in Indonesia and the Philippines. Further, those who remain active at work are mostly informal workers and are self-employed in rural areas. These are also old people who are often not eligible for an adequate pension thus leaving them with limited options to retire (Kikkawa & Gaspar, 2022).

Tables 4.3, 4.4, 4.5, 4.6 and 4.7 show the demographic profile of 47 Asian Countries, divided into three groups: ages 0–14, 15–64, and 65 and above using World Bank data (n.d.). This breakdown is critical because it reveals the relative size of each generation within a country, which has significant implications for managing a multigenerational workforce. The countries are further categorised into distinct groups: young society, aging society, aged society, and superaged society based on World Health Organization (WHO) and United Nations (UN) definitions (as quoted by Asian Development Bank Institute, 2021). '*Aging society*' is defined as one where aged 65 years or older comprise more than 7% of the population; '*aged society*' has more than 14% of the population who are older; and '*superaged society*' is a society where more than 21% is aged 65 and above. Conversely, a country where the population aged 65 and above accounts for less than 7% of the total is considered a '*young society*'. Moreover, the countries are arranged based on geographical regions in Asia to provide a clearer and more meaningful presentation of the demographic profile. This is necessary to demonstrate demographic similarities and differences within specific regions.

Table 4.3 Demographic status of countries in Central Asia, 2022 (World Bank Data (n.d.) and author's calculations)

Country name	Total population (2022)	Population aged 0–14	%	Population aged 15–64	%	Population aged 65 and above	%	Category
Afghanistan	41,128,771	17,739,100	43.13	22,404,909	54.48	984,763	2.39	Young society
Kazakhstan	19,621,972	5,821,006	29.67	12,222,736	62.29	1,578,231	8.04	Aging society
Kyrgyz Republic	6,974,900	2,402,579	34.45	4,255,513	61.01	316,808	4.54	Young society
Tajikistan	9,952,787	3,609,833	36.27	5,997,363	60.26	345,591	3.47	Young society
Turkmenistan	6,430,770	2,001,238	31.12	4,098,501	63.73	331,031	5.15	Young society
Uzbekistan	35,648,100	10,769,768	30.21	23,047,260	64.65	1,831,073	5.14	Young society

The demographics of Central Asia are predominantly defined by young societies as shown in Table 4.3. Of the six countries in the region, five (Afghanistan, Kyrgyz Republic, Tajikistan, Turkmenistan, and Uzbekistan) are classified as having young societies; that is, 83.33% of the population in these countries is under the age of 65. Therefore, since the focus is on youth, it would require not only education and skill up-gradation policies but also clean job creation for the expanding labour force. Kazakhstan is the only country that has been classified as an aging society, and it accounts for 16.67% of the region. Kazakhstan's workforce management practices must address certain unique challenges in managing an aged demographic, involving retirement planning and knowledge transfer.

Table 4.4 presents East Asia as a rather diverse picture from the point of view of demographic trends, albeit many countries belong to the aging and aged categories. Of the seven countries, three (China, North Korea, and Macao SAR) were aging societies; two (Hong Kong SAR and South Korea) were aged societies; and Japan was a superaged society. This means that East Asian countries, therefore, have 42.86% of their members aging, 28.57% are aged and 14.29% are superaged. To respond to the challenges posed by an older workforce, these countries will need policies that address how to extend working lives, improve the availability of healthcare benefits, and make workplaces more age-inclusive. Mongolia, on the other hand, is the only young society in the region. With 95.39% of its population under 65, Mongolia may focus more on educational initiatives and youth employment programs.

Table 4.4 Demographic status of countries in East Asia, 2022 (World Bank Data (n.d.) and author's calculations)

Country name	Total population (2022)	Population aged 0–14	%	Population aged 15–64	%	Population aged 65 and above	%	Category
China	1,412,175,000	243,552,723	17.25	974,838,887	69.03	193,783,391	13.72	Aging society
Hong Kong SAR, China	7,346,100	884,516	12.04	4,957,645	67.49	1,503,940	20.47	Aged society
Japan	125,124,989	14,542,937	11.62	73,138,948	58.45	37,443,104	29.92	Superaged society
Korea, Dem. People's Republic (North Korea)	26,069,416	4,926,584	18.90	18,088,961	69.39	3,053,871	11.71	Aging society
Korea, Republic (South Korea)	51,628,117	5,974,221	11.57	36,623,552	70.94	9,030,344	17.49	Aged society
Macao SAR, China	695,168	103,011	14.82	501,815	72.19	90,342	13.00	Aging society
Mongolia	3,398,366	1,103,284	32.47	2,138,395	62.92	156,687	4.61	Young society

It is important to note that one country, which is Japan, is already considered superaged since 2013 as it has 29.92% of population who are aged at least 65 years. It was tagged as the oldest large country by then (Leatherby, 2023). In 2022, almost 40% of business (which is two times higher than a decade earlier) let employees work until 70 years old and even beyond. Many Japanese companies seem to have little choice with the decision because of their currency's weakness and it was harder to hire foreign workers. The drawback is that older workers are prone to on-the-job accidents, so employers need to take some effort to ensure the safety of the employees (Matsui, 2023).

Table 4.5 shows that South Asia's demography is heavily skewed towards youth. Of the seven countries in this region, six of them (Bangladesh, Bhutan, India, Maldives, Nepal, and Pakistan) are classified as young societies making up 85.71% of total. Education, vocational training, and job creation should be the top priorities of these countries to assimilate large numbers of young people into their respective labour markets. Sri Lanka is the only country being classified under an aging society in the region, having a share of 14.29%. It needs to manage a fine line between providing for its young demographic, while also dealing with the complexities an ageing workforce poses.

Table 4.6 shows that seven of the eleven countries in Southeast Asia—the Philippines, Brunei, Cambodia, Indonesia, Laos, Myanmar, and Timor Leste—are classified as young societies. They account for roughly 63.64% of the region. These countries must prioritise efforts in education and employment for the younger populations. On the other hand, Malaysia and Vietnam are classified as aging societies, representing 18.18% of the region. In addition, Singapore and Thailand are being termed as aged societies, which accounts for 18.18% of the region. Given the mix of age groups, countries will need to formulate policies that cater to both the young and old workforce, provide flexible working conditions and proper health care support.

Demographic composition in Western Asia, as shown in Table 4.7, is predominantly youthful. Of the 16 countries in this region, nine belong to the category of young societies: Iraq, Jordan, Kuwait, Oman, Qatar, Saudi Arabia, Syria, UAE, and Yemen—representing 56.25% of the region. For these countries, youth employment along with educational development should be one of the prime concerns. Meanwhile, five countries—Armenia, Azerbaijan, Iran, Israel, and Lebanon—are recognized as societies facing aging, representing 31.25% of the region. Two countries, Cyprus and Georgia, are classified as aged societies and account for 12.5% of the region. These countries will inevitably require specific policies for their older population, which includes the training and supply of their retirement to elder care services.

Tables 4.3, 4.4, 4.5, 4.6, and 4.7 present a more data-arranged overall profile of the demographic status across different regions in Asia. They were classified as young societies, aging societies, aged societies, and superaged societies. As shown in the summary in Table 4.8, most of the Asian countries have young societies, meaning less than 7% of the countries' population is 65 years and above. In this category, 28 out of 47 which is 59.57%—fall into this classification. South Asia has the highest proportion of young societies, at 85.71%, comprising of 6 out of 7 countries. Similarly, Central Asia accounts for 83.33% (5 out of 6 countries), Southeast Asia 63.64%

Table 4.5 Demographic status of countries in South Asia, 2022 (World Bank Data (n.d.) and author's calculations)

Country name	Total population (2022)	Population aged 0–14	%	Population aged 15–64	%	Population aged 65 and above	%	Category
Bangladesh	171,186,372	44,455,660	25.97	116,383,491	67.99	10,347,220	6.04	Young society
Bhutan	782,455	173,260	22.14	560,261	71.60	48,934	6.25	Young society
India	1,417,173,173	358,634,249	25.31	960,804,385	67.80	97,734,540	6.90	Young society
Maldives	523,787	113,553	21.68	385,180	73.54	25,054	4.78	Young society
Nepal	30,547,580	8,823,440	28.88	19,864,639	65.03	1,859,500	6.09	Young society
Pakistan	235,824,862	86,230,641	36.57	139,515,778	59.16	10,078,443	4.27	Young society
Sri Lanka	22,181,000	5,061,219	22.82	14,560,801	65.65	2,558,980	11.54	Aging society

Table 4.6 Demographic status of countries in Southeast Asia, 2022 (World Bank Data (n.d.) and author's calculations)

Country name	Total population (2022)	Population aged 0–14	%	Population aged 15–64	%	Population aged 65 and above	%	Category
Brunei Darussalam	449,002	99,030	22.06	322,264	71.77	27,708	6.17	Young society
Cambodia	16,767,842	4,854,064	28.95	10,940,260	65.25	973,517	5.81	Young society
Indonesia	275,501,339	69,444,207	25.21	187,165,550	67.94	18,891,582	6.86	Young society
Lao PDR (Laos)	7,529,475	2,307,198	30.64	4,886,857	64.90	335,419	4.45	Young society
Malaysia	33,938,221	7,694,765	22.67	23,697,233	69.82	2,546,223	7.50	Aging society
Myanmar	54,179,306	13,350,169	24.64	37,136,469	68.54	3,692,668	6.82	Young society
Philippines	115,559,009	35,056,160	30.34	74,220,181	64.23	6,282,669	5.44	Young society
Singapore	5,637,022	667,397	11.84	4,117,459	73.04	852,166	15.12	Aged society
Thailand	71,697,030	11,127,205	15.52	49,663,033	69.27	10,906,792	15.21	Aged society
Timor-Leste	1,341,296	465,591	34.71	805,871	60.08	69,834	5.21	Young society
Viet Nam	98,186,856	21,984,171	22.39	67,251,030	68.49	8,951,655	9.12	Aging society

(7 out of 11 countries), and Western Asia 56.25% (9 out of 16 countries). East Asia is the one with the smallest representation, with only 14.29% (1 out of 7 countries) forming part of the societies classified as young.

Aging societies are those countries where more than 7% of the population are aged 65 years or older. This makes up 25.53% of all the countries of the world, with this category having 12 out of 47. The largest share of its nations in this grouping is East Asia, which comes at 42.89%; that is 3 out of 7 countries, and then follows Western Asia at 31.25%, or 5 out of 16 countries. In Southeast Asia, 2 out of 11 countries, or 18.18% of the countries, can be classified as aging societies; Central and South Asia each has one aging society, representing 16.67% and 14.29%, respectively.

Aged societies represent 12.77% (6 out of 47) of the countries. For East and Southeast Asia, 28.57% (2 out of 7) and 18.18% (2 out of 11) of the countries are aged, respectively. Western Asia also comprises 12.5% (2 out of 16), and both Central Asia and South Asia do not have an aged society. However, the percentage of superaged societies, where more than 21% of the population is 65 or older, in Asia is surprisingly low at only 2.13%, taking 1 out of 47 countries into consideration. Japan which belongs to East Asia, is the only Asian country that has been categorised as a superaged society.

Table 4.7 Demographic status of countries in Western Asia, 2022 (World Bank Data (n.d.) and author's calculations)

Country name	Total population (2022)	Population aged 0–14	%	Population aged 15–64	%	Population aged 65 and above	%	Category
Armenia	2,780,469	569,378	20.48	1,845,507	66.37	365,584	13.15	Aging society
Azerbaijan	10,141,756	2,381,310	23.48	7,039,605	69.41	720,841	7.11	Aging society
Cyprus	1,251,488	199,351	15.93	866,553	69.24	185,584	14.83	Aged society
Georgia	3,712,502	789,432	21.26	2,380,726	64.13	542,343	14.61	Aged society
Iran, Islamic Republic	88,550,570	20,867,571	23.57	60,933,022	68.81	6,749,976	7.62	Aging society
Iraq	44,496,122	16,790,798	37.74	26,190,212	58.86	1,515,112	3.41	Young society
Israel	9,557,500	2,682,711	28.07	5,723,598	59.89	1,151,191	12.04	Aging society
Jordan	11,285,869	3,622,111	32.09	7,230,556	64.07	433,202	3.84	Young society
Kuwait	4,268,873	884,303	20.72	3,174,078	74.35	210,493	4.93	Young society
Lebanon	5,489,739	1,506,482	27.44	3,440,173	62.67	543,084	9.89	Aging society
Oman	4,576,298	1,236,831	27.03	3,213,224	70.21	126,243	2.76	Young society
Qatar	2,695,122	426,171	15.81	2,227,982	82.67	40,970	1.52	Young society
Saudi Arabia	36,408,820	9,446,950	25.95	25,938,738	71.24	1,023,133	2.81	Young society
Syrian Arab Republic (Syria)	22,125,249	6,950,650	31.42	14,139,080	63.90	1,035,520	4.68	Young society
United Arab Emirates (UAE)	9,441,129	1,438,088	15.23	7,830,368	82.94	172,674	1.83	Young society
Yemen, Republic	33,696,614	13,309,233	39.50	19,490,300	57.84	897,081	2.66	Young society

Table 4.8 Summary of Asia's demographic status (World Bank Data (n.d.) and author's calculations)

	Central Asia	East Asia	South Asia	Southeast Asia	Western Asia	Total	Percentage (%)
Young society	5 (83.33%)	1 (14.29%)	6 (85.71%)	7 (63.64%)	9 (56.25%)	28	59.57
Aging society	1 (16.67%)	3 (42.89%)	1 (14.29%)	2 (18.18%)	5 (31.25%)	12	25.53
Aged society		2 (28.57%)		2 (18.18%)	2 (12.5%)	6	12.77
Superaged society		1 (14.29%)				1	2.13
Total	6	7	7	11	16	47	100.00

As such, strategies will need to grapple with education and youth employment in societies that have a young population. In this respect, societies in East and Western Asia should be taken as aging and aged, meaning that policies must face healthcare, retirement, and sustainability of the workforce. In this light, demographic insight is a core variable in ensuring that approaches to handling the multigenerational workforce of Asia are tailored accordingly. Societies with youthful populations, for example, may require much focus on education and initiatives for youth employment, while those in which the population is ageing may need much focus on health care, life plans, and fostering a sustainable experienced workforce. The reality of such varying demographics across Asia would mandate a very careful understanding of the challenges and opportunities of a multigenerational workforce. This is the reason it is important to understand this, making the design of impactful strategies more relevant.

4.4 Discussion of Implications

The differences between the generations especially between the older and younger ones in terms of work styles and preferred communication styles, leadership styles, and rewards are very evident that conflict may be inevitable. The potential clash, miscommunication, and misunderstanding require managers to adopt inclusive approaches in the organisation. Catering to the needs of diverse employees is challenging but definitely attainable.

Singh (2023) recommends two strategies on managing the different generations in the organisation. The first one is to accept that these stereotypes and prejudices about the generations are often blinders and restrict us from understanding team members' skills, values, and commonalities. Doria and De Vera (2019) also support this as they conclude that human behaviour in organisation is influenced not only by age or generation but by personality, job satisfaction, motivation, leadership style, individual experiences, and situational factors. Robbins and Judge (2023) agrees that managers including employees should recognize that these are just stereotypes and should not be perpetuated. It is also recommended that everyone should challenge its own stereotypical beliefs when treating employees to increase objectivity. The second strategy from Singh (2023) is to promote '*reverse mentoring*' where younger employees can mentor their senior colleagues on new age skills such as digital literacy and artificial intelligence. This is based on the concept of '*mutual learning*' which assumes that every generation has unique skills that can be shared to others, challenging the traditional belief that only older employees can mentor younger colleagues.

PricewaterhouseCoopers International Limited (PwC, 2023) conducted the Asia Pacific Workforce Hopes and Fears Survey and was published in June 2023. There were about 19,500 respondents from the region and 50% of which are millennials, 27% are Gen X, 12% are Gen Z and 10% are baby boomers. Similar to the 2022 results, only 50% of the workers in Asia Pacific feel comfortable about being themselves at work so this implies that there is room for employers to cultivate a more inclusive workplace culture that celebrates diversity and improves belonging and acceptance among workers. Employers should strive to nurture a work environment that promotes a sense of belongingness since this is a critical element towards higher retention rate and better performance. Results also show that a significant number in the workforce feels underutilised which suggests that employers should create strategies to understand the employees' aspirations, listen to their sentiments, provide better opportunities to excel, and empower the employees. It was also found out that half of employees in Asia Pacific feel disconnected from their company values. PwC recommends leaders to conduct culture assessment, even a simple one, to gain meaningful insights from the employees and plan for a refreshed approach depending on the findings. It is also recommended that leaders bring people along on the journey. It is inevitable that employees resist change even for the better if they are unsure of the future so letting employees participate in decision making such as goal setting will make them feel valued.

Since various studies recommend the most common preference across generations. It really pays to consider this and move forward from the traditional management approaches. O'Boyle (2021) found out that all generations of employees want ethical leaders. Ethical behaviour is usually associated by older generations with personal character while ethical behaviour is defined by younger generations as concern with people and the planet. Furthermore, employees know that their organisations also care about their well-being, whether it may be physically or mentally. Most organisations nowadays have wellness programs. But with the current generations, physical wellness alone isn't enough. Programs that promote better mental health should be incorporated in human resource management plans of organisations. The millennial generation desires for leaders who are open and transparent. Gen Z and younger millennials desire leaders who fosters a diverse and inclusive workplace. Diversity, equity and inclusion (DEI) efforts of organisation is not just a "nice to have" for this generation; for them it is a requirement and considered core to their personal identities.

The best managers individualise their approach. Walters (2020) recommends that managers directly ask the employees especially the millennials and generation Z about their strengths, how they liked to be managed or the things they value in the workplace. He recommends applying the '*Platinum Rule*' which is the reverse Golden Rule. Managers should treat the employees the way they want to be treated, not the way the managers themselves want to be treated.

One of the most challenging tasks of managers today is providing sufficient feedback so employees know they're progressing while establishing realistic expectations about how long it takes to advance based on performance. Being exposed to a diverse set of experiences is vital for the success of young workers. This could include working with seasoned colleagues, informal mentors, and coaches, as well as getting involved with different departments and levels within the organisation. These interactions can broaden their horizons and open doors to new possibilities they might not have known existed (Pendell & Van der Helm, 2022).

Regardless of their age or generation where they belong, Lennon (2015) recommends that the best way to manage the employees is to create an open environment where they can express themselves. The feeling of being valued comes from dialogues not policy as she emphasised. Another tip is to equip leaders with a broad range of leadership styles. Different situations call for different styles.

Robbins and Judge (2023) also advise managers that no matter what composition the team, differences can be used to improve the overall effectiveness of the group. However, both leaders and team members should be aware of '*faultlines*'. These are divisive differences that could divide the team, especially during stressful situations. To overcome these effects, teams should find the common ground among members and highlight these similarities. They should also use the different strengths that each person possesses to the group. Having a common goal or vision will help in keeping everyone focused on the same objective. However, the effective management of a multi-generational workforce is an ongoing process. Challenges and opportunities will emerge as the workforce composition evolves.

4.5 Future Directions

There is an upcoming generation that few people have heard of and that is Generation Alpha coined by Mark Crindle, a generational researcher. As baby boomers exit, this generation will also most likely change the dynamics of the workforce with their distinct characteristics. Britannica (n.d.) use this term to refer to the generation born (and who will be born) between 2010 and 2025. Again, there is no consensus for the age range of this generation so to avoid overlapping of age based on Table 4.1, their birth years will be from 2013 to 2025.

Generation Alpha are the children of the Millennials, and often the younger siblings of Generation Z. This generation can be difficult to describe demographically because they are still being born. But McCrindle, the company that introduced this term, attempts to give a background based on the preceding generations. Generations are influenced by the milestones they undergo in life and are also influenced by their grandparents, parents, and their siblings. Generation Alpha basically represents the future and serve as a lens of how we can look at the future decade. This generation is about 10 years old as of 2024, populating grade schools and the majority of this generation are still unborn. Generation Alpha is the cohort where screens are placed in front of them from the youngest age as pacifiers, entertainers, and educational aids. Alphas are the first generation fully born in this new millennium and so they are the first to be fully shaped in a global culture (McCrindle, 2022). It is likely that Generation Alpha will grow up to be creative and will think outside the box. As they mature, they will likely crave for achievement, independence, and recognition, but they may also show traits such as competitiveness, narcissism, intolerance of uncertainty, impulsiveness, a desire for attention, and a willingness to take risks. With the right upbringing, it is important to instil in them the right morals and values that match with what society expects and respects tradition. It is crucial to teach them how to handle emotions, build strong connections with others, and strike a healthy balance even when faced with technology (Jha, 2020).

Generation Alpha has yet to enter the workforce, but their future involvement presents unique challenges and opportunities for organisations. Understanding their specific traits and anticipated societal roles is essential for effective management in a multigenerational workforce. This generation is expected to build upon the collective experiences of previous generations, addressing long-term issues such as pandemic-related health concerns and societal wealth disparities. They are likely to play significant roles in promoting diversity, equity, and inclusion, and influencing legal reforms for equal pay and higher minimum wages (Sessoms-Penny, 2022). As the job market evolves, Generation Alpha will likely pioneer careers that are currently non-existent, driven by their advanced technological skills and ability to work remotely from any location (Sessoms-Penny, 2022). However, managing this generation will be challenging due to their need for constant change and adaptability (Karthikeyan & Karthikeyan, 2022). Researchers predict that Generation Alpha will delay major life milestones like marriage and retirement, resulting in a more dynamic and extended career lifespan. By 2030, McCrindle said that they are expected to make up 11% of the global workforce, distinguished by high levels of formal education, technological fluency, and substantial wealth (Sheetal et al., 2024). Strategic and collaborative planning tailored to these characteristics will be crucial for organisations to effectively integrate Generation Alpha into a multigenerational workforce. This poses an array of very important questions to organisational leaders. How can digitally native Generation Alphas be attracted and retained with their unique needs and preferences? What are the kinds of leadership styles and workplace structures are needed to optimise collaboration and active multi-generational workforce potential? These are some of the challenging issues that point out the importance of developing proper strategies in management for the next era in Asia.

4.6 Conclusion

Good leaders bring out the best in their people. The presence of a multigenerational workforce calls for new management approaches and strategies that should formally be incorporated in organisation's plans. Managers are recommended to use not just one leadership style but should be readily able to switch varying styles depending on the situation. This is the contingency style of leadership. Employees, even those belonging to the same generations, have individual differences and these differences warrant different approaches. Generational diversity is here to stay so managers should really make an effort to celebrate it instead of resisting it. As for the rewards, the generations in the workplace have different preferences thus it pays to have flexible benefit plans for them. Though this chapter presented the different common traits of the generations, it should be acknowledged that these are still 'stereotypes'. Millennials may possess characteristics of baby boomers and Generation X and vice versa. Younger generations which will comprise the majority of the workforce in the near future should be given attention since it is a challenge to keep them. The goal is to reduce turnover rates and increase job satisfaction among employees since these equate to better performance.

The emergence of Generation Alpha will reshape the future of work in Asia, as they will most likely have distinct work preferences compared to those of preceding generations. This generational shift will open new challenges and opportunities with their entry into the workforce. Being products of the digital age, their IT capabilities may be more innovative. Since this generation is adaptable by nature, the same is expected for their future managers who will need to adopt strategies that embrace their unique preferences and traits.

Managers may include deploying collaborative strategies such as continuous learning and development, flexible working arrangements, transparent culture, and open communication across generations. With these strategies, organisations can maximize the potential of a diverse workforce and create an environment where every generation thrives.

Highlights

- Each generation of workforce has distinct values and work styles
- Generation Alpha will transform Asia's future work landscape and dynamics with their distinct work preferences
- Differences between the generations require managers to adopt inclusive approaches in the workplace
- Future-ready managers individualise their approach for Asia's evolving workforce.

References

Adkins, A. (2023, July 21). Millennials: The job-hopping generation. *Gallup*. https://www.gallup.com/workplace/231587/millennials-job-hopping-generation.aspx

Anderson, H. J., Baur, J. E., Griffith, J. A., & Buckley, M. R. (2017). What works for you may not work for (Gen)Me: Limitations of present leadership theories for the new generation. *The Leadership Quarterly, 28*(1), 245–260. https://doi.org/10.1016/j.leaqua.2016.08.001

Asian Development Bank Institute. (2021). In S. C. Park, N. Ogawa, C. J. Kim, P. Sirivunnabood, & T. H. Lee (Eds.), *Demographic transition and its impacts in Asia and Europe*. https://www.adb.org/sites/default/files/publication/738926/demographic-transition-web-ready.pdf

Beresford Research. (2024). Age range by generation: Generations defined by name, birth year, and ages in 2024. https://www.beresfordresearch.com/age-range-by-generation

Bertsch, A., Saeed, M., Ondracek, J., Abdullah, A. B. M., Pizzo, J., Dahl, J., Scheschuk, S., Moore, W., & Youngren, D. (2022, March 29). Variation in preferred leadership styles across generations. *Journal of Leadership in Organizations, 4*(1). https://doi.org/10.22146/jlo.70057

Britannica. (n.d.). Generation Alpha. In S. Elridge (Writer), *The Encyclopedia Britannica*. https://www.britannica.com/topic/Generation-Alpha

Britannica. (n.d.). Subsistence farming. In A. Augustyn (Ed.). https://www.britannica.com/topic/subsistence-farming

California Employers Association. (2023, May 24). Generations and benefit packages in the workplace. https://www.employers.org/news/2023/05/24/newsletter/generations-benefit-packages-in-the-workplace/

Doria, S., & De Vera, L. (2019). A millennial leader's behavior in the organization: A descriptive case study. *Southeast Asian Journal of Science and Technology, 4*(1). https://www.sajst.org/online/index.php/sajst/article/view/196

Flavia, A. (2019). In the digital world, all roads lead to Rome. But is Rome prepared? *Dental Press Journal of Orthodontics, 24*(6), 7–8. https://doi.org/10.1590/2177-6709.24.6.007-008.edt

Jaakkola, E. (2020). Designing conceptual articles: Four approaches. *AMS Review, 10*(1), 18–26. https://doi.org/10.1007/s13162-020-00161-0

Jha, A. K. (2020, June). Understanding generation alpha [Preprint]. OSF Preprints. https://doi.org/10.31219/osf.io/d2e8g

Karthikeyan, C., & Karthikeyan, C. (2022). Exploring Generation Alpha with an HR perspective. *Human Resource Management International, 9*, 749–760.

Kikkawa, A., & Gaspar, R. (2022, September 19). Your questions answered: How can Asia manage its growing population of older workers? *Asian Development Blog*. https://blogs.adb.org/blog/your-questions-answered-how-can-asia-manage-its-growing-population-older-workers

Kim, A., Mclerney, P., Smith, T. R., & Yamakawa, N. (2020, June 29). What makes Asia-Pacific's Generation Z different? *McKinsey and Company*. https://www.mckinsey.com/capabilities/growth-marketing-and-sales/our-insights/what-makes-asia-pacifics-generation-z-different

Leatherby, L. (2023, July 16). How a vast demographic shift will reshape the word. *The New York Times*. https://www.nytimes.com/interactive/2023/07/16/world/world-demographics.html

Lennon, T. (2015). Managing a multi-generational workforce: The myths vs. The realities. *Hay Group*. https://focus.kornferry.com/wp-content/uploads/2015/02/HayGroup_Managing_multi-gen_workforce.pdf

Matsui, M. (2023, August 13). Nearly 40% of Japanese companies hire people over 70 years old. *Nikkei Asia*. https://asia.nikkei.com/Spotlight/Datawatch/Nearly-40-of-Japanese-companies-hire-people-over-70-years-old

McCrindle, M. (2022, September 15). Four distinctives of Generation Alpha. *McCrindle*. https://mccrindle.com.au/article/topic/generation-alpha/four-distinctives-of-generation-alpha/

McCrindle, M., & Wolfinger, E. (2020). Generations defined. *Ethos, 18*(1). https://www.researchgate.net/publication/306189905_Generations_defined

O'Boyle, E. (2021, March 30). 4 things Gen Z and Millennials expect from their workplace. *Gallup*. https://www.gallup.com/workplace/336275/things-gen-millennials-expect-workplace.aspx

Obsidian HR. (2024, January 31). What employee benefits each generation wants. *Obsidian HR.* https://www.obsidianhr.com/what-employee-benefits-each-generation-wants/

Pendell, R., & Vander Helm, S. (2022, November 11). Generation disconnected: Data on Gen Z in the workplace. *Gallup.* https://www.gallup.com/workplace/404693/generation-disconnected-data-gen-workplace.aspx

PricewaterhouseCoopers (PwC). (2023, June). Is the workforce ready for reinvention? Asia Pacific workforce hopes and fears survey 2023. *PwC Asia Pacific.* https://www.pwc.com/gx/en/asia-pacific/hope-and-fears/2023/asia-pacific-hopes-and-fears-survey.pdf

Purdue Global. (n.d.) *Generational differences in the workplace* (Infographic). https://www.purdueglobal.edu/education-partnerships/generational-workforce-differences-infographic/

Rathi, S., & Kumar, P. (2023). Differences in work value, communication style, and leadership style among generational cohorts at the workplace. In *Advances in human resources management and organizational development* (pp. 233–254). https://doi.org/10.4018/978-1-6684-6602-5.ch010

Robbins, S. P., & Judge, T.A. (2023). *Organizational behavior* (19th ed.) Pearson Education Limited.

Sessoms-Penny, S. G. (2022). *Collaborative and strategic planning to meet the needs of a multi-generational workforce and the organization. The Phoenix Scholar™,* 5(1), 22–27. University of Phoenix. https://www.phoenix.edu/content/dam/edu/career-institute/doc/white-paper-generations-sessoms-penny.pdf

Sheetal, S. K., Agarwal, C., & Kumar, S. (2024). HR strategies in the context of millennial and Alpha generation. *Rabindra Bharati University Journal of Economics, 17,* 90–95.

Singh, V. (2023, June 12). Managing the multigenerational workforce: Strategies for success. *Economic Times HRWorld India.* https://hr.economictimes.indiatimes.com/news/workplace-4-0/talent-management/managing-the-multigenerational-workforce-strategies-for-success/100942812

United Nations DESA (Department of Economic and Social Affairs). (2019). *World population ageing* 2019: *Highlights.* United Nations DESA Population Division. https://digitallibrary.un.org/record/3846855/files/WorldPopulationAgeing2019-Highlights.pdf

United Nations DESA (Department of Economic and Social Affairs). (2023). *World population ageing 2023: Challenges and opportunities of population ageing in the least developed countries.* UN DESA Population Division. https://desapublications.un.org/file/20355/download

United Nations. (2001). *Chapter V: Generations* (pp. 68–75). https://www.un.org/esa/socdev/rwss/docs/2001/5%20Generations.pdf

United Nations. (2024). Total population, both sexes combined (thousands): Asia and World, 1 July 2023. UNdata. https://data.un.org/Data.aspx?d=PopDiv&f=variableID%3A12

United Nations. (n.d.). *Shifting demographics.* United Nations. https://www.un.org/en/un75/shifting-demographics

Walters, P. (2020, August 26). Create a culture that inspires: Generational differences at work. *Gallup.* https://www.gallup.com/cliftonstrengths/en/318161/create-culture-that-inspires-generational-differences-work.aspx

World Bank Data. (n.d.). *Population ages 0–14, total.* https://data.WorldBank.org/indicator/SP.POP.0014.TO

World Bank Data. (n.d.). *Population ages 15–64, total.* https://data.WorldBank.org/indicator/SP.POP.1564.TO

World Bank Data. (n.d.). *Population ages 65 and above, total.* https://data.WorldBank.org/indicator/SP.POP.65UP.TO

World Bank Data. (n.d.). *Population total: 2023.* https://data.WorldBank.org/indicator/SP.POP.TOTL

Sheena D. C. Doria-De Vera is an Assistant Professor at the Institute of Management, University of the Philippines Baguio. She earned her master's degree in Management and bachelor's degree in Management Economics from the same university. Her research interests include management, organisational behaviour, managerial economics, marketing, and human resource management. Specifically, she has a strong interest in leadership, business models, entrepreneurship, generational studies, contingency valuation, industry analysis, and consumer behaviour.

Chapter 5
Mindful Futures: Navigating Mental Well-Being in the Evolving Work Environment

Clusters of Future Studies: Humanity at the Limen

Kiat Sing Heng, Chia Hua Sim, and Mei Yuan Law

Abstract This conceptual chapter examines mental well-being and work environment within the Asian context. It explores how countries navigate workplace mental health amidst post-COVID transformations, balancing traditional work values with emerging work dynamics. The chapter highlights the growing concerns surrounding employee mental well-being and the challenges of reconciling rapid technological shifts, remote work, and workplace expectations with mental wellness. Drawing insights from some Asian countries, this chapter examines policies, programs, and interventions to safeguard employee mental well-being. Additionally, it highlights the crucial role of leadership in fostering mentally healthy work environments. The comparative analysis identifies shared challenges and country-specific approaches, offering key takeaways on the importance of proactive leadership, adaptive workplace strategies, and sustainable mental health initiatives. This chapter provides a forward-looking perspective on mental well-being in Asian workplaces, emphasising actionable strategies for organisations to cultivate a resilient and mentally healthy workforce amid ongoing shifts in work culture.

Keywords Workplace mental health · Workplace culture · Well-being · Next-Gen workforce · Leadership · Future work environment

K. S. Heng (✉) · C. H. Sim · M. Y. Law
Swinburne University of Technology Sarawak Campus, Kuching, Malaysia
e-mail: kheng@swinburne.edu.my

C. H. Sim
e-mail: chsim@swinburne.edu.my

M. Y. Law
e-mail: mlaw@swinburne.edu.my

C. Y. Fung et al. (eds.), *Future of Work in Asia*,
https://doi.org/10.1007/978-3-031-95804-5_5

5.1 Introduction

5.1.1 The Background

Mental health, as defined by the World Health Organization (WHO), is a condition of optimal physical and mental functioning that empowers individuals to engage in productive work and make positive contributions to society (World Health Organization, 2005). WHO subsequently emphasises the significance of mental health in community engagement and the management of life's stresses (World Health Organization, 2022). Historically, the workforce was dominated by manual labour due to technological constraints, leading to minimal automation. This era featured job security from loyalty but limited career progression and a lack of work-life balance (Lim, 2023). In contrary, today's workforce prioritises continuous learning, skill development, work-life balance, and mental health, reflecting a shift towards holistic well-being in modern work environments (Lim, 2023).

Specifically, WHO highlights significant mental health disparities, with 75% of those affected residing in Low- and Middle-Income Countries (LMICs). In these regions, stigma, lack of awareness, limited resources, and inadequate research intensify the marginalisation of mental health services, leading to reduced care-seeking and access to evidence-based treatments (Sweetland et al., 2014). As countries navigate shifts in work culture, the critical need for mental wellness in the workplace is often overlooked. Organisations typically offer superficial solutions, such as Employee Assistance Programs and limited mental health services in benefits packages, a minimalist approach criticised for not fully addressing workplace mental health's complex needs (Kelloway et al., 2023).

More recently, societal and work evolution, especially during the COVID-19 pandemic, has forced organisations to adjust management practices, affecting communication, performance evaluation, and workload distribution (Oakman et al., 2020). Burnout, limited community support, and fear of losing work-life control show the pandemic's ongoing impact on workplace dynamics (Campbell & Gavett, 2021). Southeast Asian countries lag behind other regions in mental health care due to nascent national-level legislation and policy development, public perceptions, stigma, and inadequate awareness of mental health conditions (Dessauvagie et al., 2022).

5.1.2 Chapter Aim

This conceptual chapter addresses the pressing issue of mental wellness in the workplace in Malaysia while drawing comparisons with other Asian countries. In an era of transformative shifts in work cultures, the significance of mental health is often overshadowed by the rapid evolution of work processes. This chapter highlights the necessity of prioritising employee well-being in the future workplace by examining how different nations navigate workplace mental wellness amidst changing socio-economic landscapes. Exploring Malaysia's journey alongside its regional counterparts provides insights into shared and country-specific challenges shaping mental health at work. Additionally, the chapter offers a forward-looking perspective on integrating mental health considerations into evolving work environments.

5.1.3 Comparative Insights on Workplace Mental Health

Globalisation has significantly transformed the workplace, and Malaysia has been no exception to this trend. As a vibrant Southeast Asian country, the workplace is changing rapidly (Yap et al., 2023). In Malaysian work culture, there has been a long-standing emphasis on collectivism, highlighting the importance of loyalty, teamwork, and organisational commitment, with some variations among the ethnic communities (Hyun & Yoan, 2018). However, the work values in Malaysia have experienced shifts due to rapid changes brought about by the recent global challenges.

The impact of the COVID-19 pandemic on workplace mental health has varied across Malaysia, Singapore, Thailand, and the Philippines, reflecting differences in economic structures, policy responses, and cultural attitudes toward work and well-being. While all four countries report increased mental health concerns, their approaches to addressing these challenges differ in effectiveness and scope.

One major factor driving these differences is the accelerated integration of Fifth Industrial Revolution (IR5.0) technologies, which has reshaped regional workplace dynamics. In Malaysia, for instance, the pandemic hastened the adoption of advanced digital tools, automation, and remote work practices (Yap et al., 2023). This technological shift has altered labour and employment patterns, particularly affecting multinational corporations that now navigate increasing operational changes and technology dependence (Khor & Tan, 2023).

As a result, Malaysia has experienced a notable transformation in work values, characterised by an increased reliance on digital technologies and a growing preference for flexible work arrangements (Amankwah, 2023). These shifts redefine how work is performed and reshape workforce expectations and priorities, influencing employees' overall work experience. However, despite these changes, the transition to remote work and greater technological dependence has not necessarily translated into improved employee well-being. Malaysia's small and medium enterprises (SMEs), a crucial pillar of the economy, have witnessed rising levels of stress, depression, and anxiety among workers, suggesting that the evolving workplace landscape presents new mental health challenges rather than alleviating existing ones (Abdullah et al., 2022; Jalil & Ali, 2023). This trend suggests that the rapid digitalisation of work may have outpaced the development of supportive mental health policies. Unlike Singapore, where organisations have implemented resilience-building strategies (Ditchburn & Koh, 2024), Malaysia appears to lack a systematic approach to mitigating workplace stress, leaving employees vulnerable to deteriorating mental health.

Singapore, in contrast, demonstrates a more structured response to workplace mental well-being. While organisational changes during the pandemic heightened work pressure and stress, the country's emphasis on resilience-building has played a crucial role in buffering these negative effects. Policies promoting work-life balance, social connectivity, and positive interpersonal relationships indicate a proactive stance in addressing mental health concerns (Ditchburn & Koh, 2024).

Meanwhile, Thailand presents a broader national-level perspective, highlighting the long-term psychological impact of COVID-19 on its workforce. The country's increased prevalence of mental health issues has highlighted the urgent need for accessible psychological support services and targeted interventions (Mahikul et al., 2024). Unlike Singapore's workplace-centred approach, Thailand's response emphasises the role of economic support and social connectivity in improving overall quality of life. This suggests that while organisational policies are crucial, mental health challenges in the workforce cannot be divorced from broader socioeconomic conditions.

Bollettino et al. (2023) provide an insightful analysis of the evolving mental health landscape in the Philippines during the COVID-19 pandemic, revealing both progress and persistent challenges. While mental health providers observed a notable increase in psychological distress among the population, they also report a decline in stigma surrounding mental health services. This suggests a shift in public perception, potentially driven by heightened awareness and widespread discussions about mental well-being during the crisis. However, despite this progress, significant barriers to accessing mental health services remain, including financial constraints, lack of institutional support, and insufficient mental health infrastructure. These findings highlight the persistent gap between increased awareness and the availability of adequate mental health interventions, raising concerns about the sustainability of mental health support beyond the pandemic.

Expanding on these challenges, Maravilla and Tan (2021) take a broader perspective by critiquing the Philippine Mental Health Act, arguing that its effectiveness is undermined by a lack of integration with economic policies. They highlight the bidirectional relationship between mental health and economic stability, emphasising that financial insecurity and inadequate funding for mental health initiatives exacerbate psychological distress. While the Mental Health Act marks progress in legislative recognition, the authors question whether it translates into tangible improvements in mental health services and economic resilience. Without structural reforms addressing economic barriers to mental healthcare, the legislation risks being a symbolic gesture rather than a transformative policy. This concern aligns with Bollettino et al.'s (2023) findings, reinforcing the idea that policy advancements must be accompanied by institutional and economic support to be effective.

Further reinforcing the importance of policy initiatives, Alibudbud's (2022) study examines public interest in mental health across Malaysia, the Philippines, Singapore, and Thailand from 2004 to 2021 using Google search trends. The findings reveal that mental health-related searches increased following the enactment of mental health legislation in Malaysia, the Philippines, and Singapore, indicating that policy initiatives play a role in raising awareness and reducing stigma. Additionally, the

COVID-19 pandemic led to a surge in mental health-related searches across all four countries, highlighting the growing concern for psychological well-being during crises. These results demonstrate the broader regional trend in which legislative actions and external stressors shape public engagement with mental health. However, as highlighted by Maravilla and Tan (2021) and Bollettino et al. (2023), awareness alone is insufficient; in the absence of structural reforms and economic support, policies risk falling short in delivering meaningful change. This interplay between awareness, policy implementation, and systemic challenges underscores the need for holistic and sustainable mental health interventions across Southeast Asia.

Apart from that, in their editorial, Tan et al. (2023) critically examine the persistent neglect of mental health in Southeast Asia. Despite the region's diverse economic landscapes, a common thread is the systemic under-prioritisation of mental health, leading to inadequate policy frameworks and insufficient resource allocation. The authors argue that this oversight has resulted in fragmented mental health services and limited access to care, exacerbating the burden of mental health disorders across these nations. The editorial highlights the urgent need for comprehensive reforms, advocating for the integration of mental health into primary healthcare systems and the development of culturally sensitive interventions.

This growing concern about mental health across Southeast Asia is further reinforced by broader regional assessments. According to the Asia Mental Health Index Report, mental health among Asian employees is under strain, showing higher risk profiles and decreased productivity compared to other regions (AON, 2023). The key findings are outlined in Fig. 5.1.

Specifically, the Mental Health Index score for Malaysia stands at 59.6, accompanied by the following details.

How Malaysian employees feel:

- 45% feel anxious

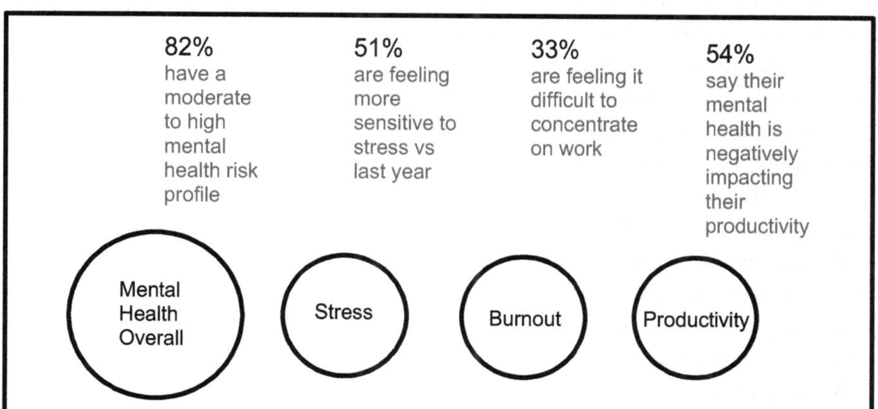

Fig. 5.1 Key findings in Asia mental health index 2023 [adapted from The Asia Mental Health Index Report (AON, 2023)]

- 44% feel depressed
- 40% feel isolated.

Hence, addressing workplace mental health is vital as Malaysia faces technological and work changes, emphasising the need to adapt to evolving work values for a healthy and productive environment.

In summary, while all four countries acknowledge the mental health challenges brought by COVID-19, their responses vary significantly. Singapore adopts an organisational approach emphasising resilience, the Philippines critiques policy gaps, Malaysia grapples with workplace structural issues, and Thailand highlights the socioeconomic determinants of mental health. These differences highlight the need for a more holistic and integrated approach that combines workplace strategies, policy reforms, and socioeconomic interventions to enhance mental well-being across different national contexts.

5.2 Key Concepts and Issues

5.2.1 Mental Well-Being in the Future of Work

This section explores the significance of mental well-being in the evolving workplace, focusing on Malaysia's corporate mental health landscape. Well-being is the integration of feeling good and functioning effectively. Feeling good involves experiencing happiness, contentment, interest, engagement, confidence, and affection, while functioning effectively entails realising potential, maintaining control over life, pursuing meaningful goals, and fostering positive relationships (Huppert, 2009). Additionally, the conceptualisation of mental well-being resonates with the definition of positive mental health (World Health Organization, 2005) discussed in the Introduction section earlier. As Malaysia navigates changes in work culture, prioritising mental health becomes crucial amid challenges like integrating technology with remote work and ensuring employee well-being. The struggle for work-life balance is intense among Malaysian professionals, who face increasing work demands and value family time. Achieving this balance is essential, as it is closely linked to positive mental health outcomes (Liswandi & Muhammad, 2023; Sarpong, 2024).

Specifically, 45% of Malaysian companies are behind the Asian average of 48% in adopting employee well-being programs. Only 35% are updating work practices for better mental health, versus Asia's 39%. Further, only 21% provide virtual mental health care, below the 26% Asian average, highlighting the need for enhanced mental health initiatives in Malaysia (BusinessToday, 2023).

Although the enforcement of the Mental Health Act 2001 in 2010 established a framework for the comprehensive care of individuals with mental disorders, it does not include specific laws or policies addressing workplace mental health. This legislative gap has contributed to a growing concern over workplace mental well-being in Malaysia (Azmi et al., 2021). Additionally, inadequate supervisor support and the pandemic's impact on employees' personal lives have further exacerbated the issue (Yusof et al., 2022). These challenges highlight the urgent need for targeted policies to safeguard and promote mental health in professional settings.

5.2.2 Challenges in the New Work Environment

IR 5.0 poses several challenges to employee well-being, mainly due to the increasing integration of advanced technologies in the workplace. Firstly, employees must acquire new skills to work alongside advanced robotics and smart machines. This requirement extends beyond soft skills to technical abilities, such as programming industrial robots and managing tasks in new job environments. The shift can be demanding and requires high technical proficiency (Adel, 2022).

Integrating IR 5.0 technologies, including customised software and artificial intelligence, demands increased employee time and effort, posing challenges in adaptation and proficiency (Adel, 2022). The vital role of technology in influencing workplace mental health is often overlooked (Johnson et al., 2020). The future of work requires substantial investment in human capital, urging workers to collaborate and adapt (Galipeau, 2023). Researchers emphasise organisations' responsibility to consider employees' mental health during this transformation (van der Walt & Lezar, 2019). With technology's growing significance, a rise in technology-related mental health concerns is anticipated (Johnson et al., 2020).

Further, the COVID-19 pandemic has disrupted the traditional work paradigm, posing challenges for employers and employees. Limited research on the psychological, social, and situational factors affecting the shift to remote work in Southeast Asia hinders the development of effective policies and management strategies for addressing issues such as fatigue, isolation, and depression (Yee et al., 2023).

Meanwhile, Zarin et al. (2022) highlight a notable gap in research related to coping mechanisms. They pointed out that while stress among teachers is a recognised issue, there is limited research focusing on how teachers cope with this stress. The scarcity of studies on coping strategies suggests a lack of understanding and resources available for teachers to manage their stress effectively. This knowledge gap can leave teachers without the necessary tools to combat stress, potentially intensifying its effects (Zarin et al., 2022).

In this regard, Andary et al. (2023) identify several key barriers impacting the access and utilisation of mental health care services in Southeast Asia. The authors noted that there is a lack of understanding and awareness about mental health issues and services also hinder access to care. Additionally, personal beliefs and attitudes towards mental health can prevent individuals from seeking help. Cultural views about mental health might sometimes prevent access to services. In essence, the challenges in Malaysia's new work era are multifaceted. They include adapting to IR5.0 technologies, addressing work-life balance issues, and navigating the complexities of remote work.

5.2.3 Safeguarding Employee Well-Being

The next part will discuss pioneering Malaysian enterprises taking innovative employee well-being practices. Some Malaysian companies have implemented innovations in mental health and workforce wellbeing.

5.2.3.1 QI Group

The QI Group's HR team in Malaysia received the 2023 Human Resources Online Employee Experience Award for Best Employee Wellness Strategy (QI Group, 2023). Their winning approach featured remote work options, virtual wellness programs, and training initiatives, prioritising mental health, work-life balance, and employee skill development in response to the challenges of the evolving work landscape.

5.2.3.2 AIA Malaysia

AIA Malaysia introduces a pioneering end-to-end mental health solution, collaborating with ThoughtFull, to integrate comprehensive mental wellness into employee benefits. The AIA Mental Health Solution includes guidance for organisations, stress management programs, and digital access to unlimited coaching, self-guided content, and a crisis hotline, covering consultations, medication, and treatment with cashless options for employees and their dependents (Insurance Asia, 2023).

5.2.3.3 Petronas

In November 2021, PETRONAS collaborated with the Malaysian Psychiatric Association (MPA) to introduce the MIND-A-CARE First Aid for Mental Health Training Programme (PETRONAS, 2022). This initiative aims to certify designated first aiders and selected employees, equipping them with skills to assist colleagues in mental distress and guide them to PETRONAS' myFriends Employee Assistance Programme. Emphasising the crucial significance of mental health, PETRONAS reaffirms its commitment to the overall well-being of its employees, particularly in challenging times (PETRONAS, 2022).

5.2.4 Comparison of Employee Well-Being Across Asia

Over the past decade, the awareness of employee well-being has accelerated in many developed Asian nations, leading to several policy-driven interventions. In Japan, for instance, the government introduced a mandatory annual stress assessment for firms with more than 50 employees under the Stress Check Program. This initiative helps employers monitor employee well-being and address psychological strain, particularly amid concerns over "karoshi," or death from overwork (Yukawa et al., 2024).

South Korea has also adopted a proactive stance by legally capping weekly working hours at 52 and targeting workplace harassment through new legislation (Han et al., 2021). Prominent corporations like Naver and Netmarble have enhanced mental health services by establishing support centres and enforcing work-life balance, including eliminating unpaid overtime (Financial Times, 2024). These national initiatives reflect a shift towards more sustainable and supportive work cultures in both countries.

In China, for instance, the Shanghai Mental Health Center and the Shanghai Pilot Health Promotion Center have collaborated to offer mental health courses for employees (China Daily, 2024). These courses, attended by nearly 300 participants, including corporate managers and HR specialists, focus on emotional management and psychological resilience, equipping employees to handle workplace stress and support colleagues better. Additionally, in December 2023, China's Ministry of Civil Affairs, the National Health Commission, and the China Disabled Persons' Federation launched a national platform to enhance mental health services (China Development Brief, 2024). This platform facilitates the sharing of patient information and streamlines referrals for community rehabilitation services, aiming to improve care for individuals with mental disorders.

In Southeast Asia, Singapore has taken structured steps to enhance employee well-being in the workplace through government-endorsed guidelines that promote mental health champions and employee counselling access. Despite such policies, program participation remains low, with under 20% of employees engaging in available support, often due to persistent stigma, privacy concerns, and doubts about the value of these initiatives (Intellect, 2024).

India's employee well-being initiatives have primarily centred on legislative reform and public access to mental health support. The Mental Health Care Act 2017 marked a turning point by decriminalising suicide attempts and mandating free mental health care and rehabilitation for affected individuals (Vadlamani & Gowda, 2019). Further, the Indian government launched KIRAN, a multilingual 24/7 mental health helpline offering psychological support and referrals (Agrawal, 2020). These interventions aim to dismantle the deep-rooted stigma surrounding mental health and provide equitable access to employee well-being support. Across these five countries—Japan, South Korea, Singapore, China, and India—efforts are gaining traction, but persistent cultural resistance, underutilisation of resources, and uneven implementation continue to pose significant challenges. Continuous improvement and cross-sector collaboration are essential to foster more inclusive, mentally healthy work environments throughout Asia.

5.3 Discussion of Implications

5.3.1 Next-Gen Workforce: Emerging Policies, Programs, and Interventions

Influenced by diverse generational experiences, the contemporary workplace is witnessing a pivotal shift as Millennials and Generation Z emerge (Nnambooze & Parumasur, 2016). This transition prompts organisations to re-evaluate strategies for motivating and optimising the performance of these cohorts. Unlike their predecessors, Millennials and Generation Z prioritise work-life balance and exhibit tech proficiency, challenging traditional workplace norms (Bulut & Maraba, 2021). Millennials comprise over 59% of the workforce globally, with Generation Z assuming significant roles (Lee, 2021). This transformative scenario necessitates organisations to thoroughly reassess and restructure their approaches to meet these emerging generations' distinctive needs and values.

Specifically, Buzza's research (2017) highlights job advancement and work-life balance as crucial considerations for Millennials in job choices, offering insights for effective recruitment strategies. Corporations can leverage these factors to attract and engage millennial talent with targeted resources by leveraging these factors. Yap and Zainal Badri's findings (2020) emphasise key contributors to millennial employees' emotional well-being, including relationships, work-life balance, autonomy, meaningful work, effective leadership, and expressions of gratitude. Due to their upbringing, millennials may be more susceptible to emotional distress (Buzza, 2017), highlighting the need for regular assessments and emotional support from managers, as recommended by Zainal Badri et al. (2020), to nurture the flourishing of millennial employees.

In essence, Becker's (2022) study offers insights with the potential to positively impact Generation Z's well-being and address mental health challenges. Despite the adversity of the COVID-19 pandemic, Generation Z displays resilience and optimism, showcasing effective coping mechanisms. Understanding these strategies allows tailored interventions and support systems to reinforce positive mental health. The study underscores the importance of face-to-face connections and mentorship options, enhancing the potential for positive outcomes. Implementing these findings can alleviate stress and anxiety related to career transitions, contributing to the overall enhancement of Generation Z's mental well-being.

Chen et al. (2023) highlight the mental health challenges Generation Z employees face in China, particularly under the demanding "996" work culture—working 9 a.m. to 9 p.m., six days a week. This unsustainable system fuels burnout and psychological distress, necessitating urgent reforms. Organisations must adopt policies promoting work-life balance, such as limiting work hours and introducing flexible schedules. Mental health interventions, including employee support programs and workplace well-being initiatives, are also crucial. A failure to address these concerns jeopardises employees' well-being and threatens organisational sustainability due to decreased productivity and high turnover rates.

Similarly, Kim et al. (2021) identify clinical experience satisfaction as a pivotal factor influencing the turnover intentions of Generation Z nursing students in Korea. Dissatisfaction with clinical placements can lead to increased stress and a diminished sense of professional fulfilment, contributing to mental health challenges and a heightened desire to leave the profession within two years of employment. This highlights the need for collaborative efforts between educational institutions and healthcare facilities to improve the quality of clinical placements. By fostering supportive learning environments, providing competent mentorship, and ensuring meaningful clinical experiences, stakeholders can enhance student satisfaction, bolster mental well-being, and reduce turnover intentions among future nurses.

Millennials and Generation Z can promote mental well-being by fostering a healthy work culture, advocating for mental health initiatives, and encouraging open workplace communication. By aligning organisational strategies with the values of these generations, workplaces can cultivate environments that support productivity and employee well-being.

5.3.2 Role of Leadership in the Future of Work

Kelloway and Drapeau (2017) advocate prevention, intervention, and accommodation for workplace mental health leadership. The model prioritises preventive measures over crisis interventions, highlighting the organisational responsibility to address workplace factors contributing to mental health issues. It also stresses the need for accommodation policies to support individuals during and after a mental health crisis.

In this critical context, leaders and team members must advocate for an understanding of the specific areas where mental health disparities are most pronounced. Furthermore, they must passionately call for a more equitable distribution of information, resources, and care (Kaslow et al., 2021). This approach underscores the vital role of leadership in combatting mental health issues by proactively addressing disparities and ensuring a fair allocation of support and resources.

On the other hand, the initial stride towards establishing a workplace environment that is both safe and conducive to well-being involves the recruitment of leaders possessing elevated emotional competencies and a strong sense of personal well-being, as argued by Haver et al. (2019). By critically assessing and prioritising such qualities in leadership recruitment, organisations can set the foundation for addressing mental health issues and fostering a workplace that values the psychological welfare of its members.

Meanwhile, leadership behaviours like support, trust, and feedback also mitigate psychosocial hazards' impact on employee mental health, promoting worker well-being and stress management (Naji et al., 2022). Specifically, empowering leadership, marked by participatory and positive attitudes in Malaysian SMEs, significantly improves their employees' mental health (Jalil et al., 2023).

Grimes et al. (2022) examine the role of leadership in fostering resilience and promoting psychologically healthy workplaces during and beyond the COVID-19 pandemic. They outline eight key leadership strategies: building personal resilience, practising compassionate leadership, ensuring authentic communication, fostering psychological safety, supporting team resilience, prioritising employee well-being, leveraging networks and communities of practice, and contributing to a collaborative national strategy. These strategies emphasise the importance of leaders not only in guiding their organisations through crisis but also in sustaining long-term psychological health within the workforce. These findings contribute to the growing discourse on leadership's influence in shaping psychologically healthy workplaces and highlight the necessity of proactive leadership in navigating current and future workplace challenges.

Leadership also plays a crucial role in shaping workplace culture and ensuring employee well-being in Japan. Dollard and Loh (2023) introduce the Psychosocial Safety Climate (PSC) concept, which refers to an organisation prioritising psychological health and safety. Their study highlights how organisations with a strong PSC experience lower workplace stress and burnout levels. Additionally, Yamamoto et al. (2024) suggest that leaders who respond effectively to workplace concerns, such as addressing near-miss incidents, help to reduce employee anxiety and improve retention.

5.4 Future Directions

Due to the growing importance of mental health in the workplace across Asia, it is becoming increasingly necessary to establish comprehensive policy frameworks and standards to safeguard and promote mental health at work. While the United Kingdom has implemented initiatives like Thriving at Work, many developing countries, such as Malaysia, have yet to adopt specific, comprehensive legislation or policies addressing mental well-being in the workplace (Azmi et al., 2021). Looking toward the future, Asian countries could benefit from adapting progressive mental health models thoughtfully aligned with their socio-cultural values, ensuring that workplace strategies are inclusive and contextually relevant.

Workplace stress remains widespread in most Asian countries, such as Malaysia, with many employees reporting work-related stress (The Star, 2022). However, access to treatment remains low, with only about 20% of individuals with mental health issues receiving care (Chua, 2020). This gap is likely to persist unless there is a national commitment to promoting awareness and ensuring adequate mental health resources and support, particularly regarding treatment accessibility (Villarreal-Zegarra et al., 2022). Research shows that awareness and usage of mental health resources are linked to reduced psychological distress among workers (Chan et al., 2021), underscoring the need for workplace interventions that prioritise work-life balance and employee wellness (Panbarasan et al., 2020).

In the context of future workplaces in Asia, technological solutions will likely become a key enabler of accessible and scalable mental health support. Integrating mental health apps, online counselling platforms, and virtual support groups can create a more inclusive mental health ecosystem (Kelloway et al., 2023). Digital interventions such as eHealth for Mental Wellness (EHMW) can be particularly effective in Asia's fast-evolving digital work environments. EHMW enables proactive tracking, communication, and early interventions, especially for remote and gig workers (Tan et al., 2022). Such approaches are timely and cost-effective, aligning well with the resource constraints and digital aspirations of many Asian economies.

Moreover, culturally sensitive community-based interventions are also gaining traction. Lyles et al. (2023) recommend interventions such as Functional Analytic Psychotherapy (FAP), caregiver involvement, and scalable solutions like the Motivational Cognitive Behavioural Intervention (MCBI), which are both cost-effective and culturally adaptable. When implemented via community outreach and lay health workers, MCBI presents a viable path to closing the mental health treatment gap not only in South Asia but also in other parts of Asia.

Looking ahead, the future of work in Asia will likely see an acceleration of flexible work arrangements, digitised workplaces, and greater emphasis on employee-centric policies. These trends respond to technological disruptions and reflect shifting generational values around work-life integration and mental well-being (Costa & Portioli-Staudacher, 2021). Malaysian organisations, for instance, are increasingly encouraged to adopt flexible schedules and remote work options to reduce work-life conflict and improve overall satisfaction (Au & Ahmed, 2015). In such settings, robust Occupational Health and Safety (OHS) policies will be critical to ensure employee well-being, including guidelines around job design, leave entitlements, and support services (Wells et al., 2023).

Regional experiences during the COVID-19 pandemic have further reinforced the importance of psychological safety in the workplace. Yang et al. (2024) illustrate how transparent communication and effective psychological support reduced anxiety among industrial workers in China. Similarly, corporate social responsibility (CSR) initiatives in China's hospitality sector have shown promise in mitigating burnout and fostering a positive work culture (Wang et al., 2025). These developments point to a broader shift in Asian workplaces from reactive crisis management to proactive and sustained investments in employee well-being. This evolving work culture necessitates empathetic, forward-looking leadership that embeds mental health and wellness as core elements of organisational sustainability and long-term success.

5.5 Conclusion

Cultivating a mentally healthy workplace requires a culture that prioritises trust, open communication, and strong leadership. This approach is essential for organisations, as they navigate the evolving nature of work alongside broader workplace trends across Asia. While some advanced Asian countries, such as China, Japan, Singapore and India, have made progress in recognising the importance of workplace mental health, disparities in policy implementation, resource allocation, and cultural attitudes persist. In the future, a multifaceted approach is needed to enhance access to mental healthcare, embedding mental health awareness into corporate policies and fostering work environments that emphasise purpose and psychological safety. By leveraging innovative strategies and learning from best practices within Asia, organisations can build a resilient workforce that thrives amid rapid transformation. Integrating mental wellness into workplace strategies is a moral imperative and a crucial driver of long-term productivity, engagement, and organisational success.

Highlights

- Mental health in Asian workplaces needs policy-driven attention.
- Low treatment access highlights the urgency for awareness and support systems.
- Culturally aligned, contextually relevant models are essential.
- Digital tools and flexible work boost accessibility.
- Leadership must shift from reactive to proactive well-being strategies.

References

Abdullah, N., Hashim, N. Z. I., Zainal Abidin, I. S., & Mohd Shazali, S. M. (2022). Impact of COVID-19 pandemic in Malaysia: A critical survey. *Pertanika Journal of Social Sciences and Humanities, 30*(4), 1857–1879. https://doi.org/10.47836/pjssh.30.4.20

Adel, A. (2022). Future of industry 5.0 in society: Human-centric solutions, challenges and prospective research areas. *Journal of Cloud Computing: Advances, Systems and Applications, 11*(1), 40–40. https://doi.org/10.1186/s13677-022-00314-5

Alibudbud, R. (2022). Insights from an infodemiological study of mental health in Malaysia, the Philippines, Singapore, and Thailand from 2004 to 2021. *Asian Journal of Psychiatry, 78*, 103311–103311. https://doi.org/10.1016/j.ajp.2022.103311

Amankwah, M. O. (2023). Pre-Covid and peri-Covid analysis of employees' work values and job quality. *African Journal of Economic and Management Studies, 14*(2), 223–237. https://doi.org/10.1108/ajems-07-2022-0308

Andary, S., Bassani, J., Burrell, G., Cole, E., Evans, R., Redman, E., & Kumar, S. (2023). Barriers and enablers to access and utilization of mental health care services across Southeast Asia: A preliminary scoping review. *Asia-Pacific Psychiatry, 15*(4), Article e12549. https://doi.org/10.1111/appy.12549

AON. (2023). *2023 Asia mental health index report*. Retrieved February 4, 2024, from https://www.aon.com/getmedia/c36a31bd-273e-446f-b0f5-6944a038f128/Aon-and-TELUS-Health-Asia-Mental-Health-Index-Report-2023.pdf

Agrawal, S. (2020, September 7). Govt launches KIRAN, a 24x7 helpline for people to seek mental health counselling. *The Print*. https://theprint.in/india/govt-launches-kiran-a-24x7-helpline-for-people-to-seek-mental-health-counselling/497460/

Au, W. C., & Ahmed, P. K. (2015). Exploring the effects of workplace support on work-life experience: A study of Malaysia. *Human Resource Development International, 18*(4), 346–365. https://doi.org/10.1080/13678868.2015.1019816

Azmi, R., Ahmad, S. N. S., & Kamil, B. A. M. (2021). Mental health well-being policy and strategy in the workplace in the United Kingdom (UK): Lessons to be learned. *Central Asia and the Caucasus, 22*(5), 844–854. https://doi.org/10.37178/ca-c.21.5.067

Becker, K. L. (2022). We want connection and we do not mean Wi-Fi: Examining the impacts of COVID-19 on Gen Z's work and employment outcomes. *Management Research Review, 45*(5), 684–699. https://doi.org/10.1108/mrr-01-2021-0052

Bollettino, V., Foo, C. Y. S., Stoddard, H., Daza, M., Sison, A. C., Teng-Calleja, M., & Vinck, P. (2023). COVID-19-related mental health challenges and opportunities perceived by mental health providers in the Philippines. *Asian Journal of Psychiatry, 84*, 103578–103578. https://doi.org/10.1016/j.ajp.2023.103578

Bulut, S., & Maraba, D. (2021). Generation Z and its perception of work through habits, motivations, expectations, preferences, and work ethics. *Psychotherapy Research Study, 4*(4), 1–5. https://doi.org/10.31031/PPRS.2021.04.000593

BusinessToday. (2023, April 20). *Employers in Malaysia prioritise skills, financial well-being more than others.* Retrieved February 4, 2024, from https://www.businesstoday.com.my/2023/04/20/employers-in-malaysia-prioritise-skills-financial-well-being-more-than-others/

Buzza, J. S. (2017). Are you living to work or working to live? What millennials want in the workplace. *Journal of Human Resources Management and Labor Studies, 5*(2), 15–20. https://doi.org/10.15640/jhrmls.v5n2a3

Campbell, M., & Gavett, G. (2021, February 10). What COVID-19 has done to our well-being, in 12 charts. *Harvard Business Review.* Retrieved February 4, 2024, from https://www.inspired-engagement.com/wp-content/uploads/What-Covid-19-Has-Done-to-Our-Well-Being-in-12-Charts.pdf

China Daily. (2024, July 29). *Stigma lifted on dealing with workplace stress.* Retrieved April 11, 2025, from https://www.chinadaily.com.cn/a/202407/29/WS66a6de9ba31095c51c5105fe.html

China Development Brief. (2024, January 22). *China launches national platform for mental health services.* Retrieved April 11, 2025, from https://chinadevelopmentbrief.org/reports/china-launches-national-platform-for-mental-health-services/

Chan, C. M. H., Ng, S. L., In, S., Wee, L. H., & Siau, C. S. (2021). Predictors of psychological distress and mental health resource utilization among employees in Malaysia. *International Journal of Environmental Research and Public Health, 18*(1), 314. https://doi.org/10.3390/ijerph18010314

Chen, X., Masukujjaman, M., Al Mamun, A., Gao, J., & Makhbul, Z. K. M. (2023). Modeling the significance of work culture on burnout, satisfaction, and psychological distress among the Gen-Z workforce in an emerging country. *Humanities & Social Sciences Communications, 10*(1), 828–912. https://doi.org/10.1057/s41599-023-02371-w

Chua, S. N. (2020). Workplace mental health: The business costs. *Relate Mental Health Malaysia.* Retrieved February 4, 2024, from https://relate.com.my/the-business-costs-of-mental-health/

Costa, F., & Portioli-Staudacher, A. (2021). Labor flexibility integration in workload control in Industry 4.0 era. *Operations Management Research, 14*(3–4), 420–433. https://doi.org/10.1007/s12063-021-00210-2

Dessauvagie, A. S., Dang, H.-M., Nguyen, T. A. T., & Groen, G. (2022). Mental health of university students in Southeastern Asia: A systematic review. *Asia Pacific Journal of Public Health, 34*(2–3), 172–181. https://doi.org/10.1177/10105395211055545

Ditchburn, G., & Koh, R. E. (2024). Workplace pressure, employee stress, mental well-being and resilience in response to COVID-19 in Singapore. *Evidence-Based HRM: A Global Forum for Empirical Scholarship, 12*(2), 441–457. https://doi.org/10.1108/EBHRM-10-2022-0252

Dollard, M., & Loh, M. Y. (2023). Psychosocial safety climate in Japanese workplaces. *Journal of Occupational Health, 65*(1), e12430. https://doi.org/10.1002/1348-9585.12430

Galipeau, D. (2023). How Industry 4.0 influences our work environment. In T. Endress (Ed.), *Digital project practice for new work and Industry 4.0* (1st ed., Vol. 1, pp. 75–98). CRC Press. https://doi.org/10.1201/9781003371397-6

Financial Times. (2024, January 15). *South Korea starts to raise game on workplace mental health.* Retrieved April 11, 2025, from https://www.ft.com/content/6e70f7bd-e311-41df-94fe-7a5575493ae6

Grimes, K., Matlow, A., Tholl, B., Dickson, G., Taylor, D., & Chan, M.-K. (2022). Leaders supporting leaders: Leaders' role in building resilience and psychologically healthy workplaces during the pandemic and beyond. *Healthcare Management Forum, 35*(4), 213–217. https://doi.org/10.1177/08404704221090l2

Han, S., Ko, Y., Moon, J. E., & Cho, Y. S. (2021). Working hours are closely associated with depressive mood and suicidal ideation in Korean adults: A nationwide cross-sectional study. *Scientific Reports, 11*(1), 23102. https://doi.org/10.1038/s41598-021-02574-8

Haver, A., Olsen, E., & Akerjordet, K. (2019). Well-being among hotel managers: A study on the influence of job stressors and cognitive reappraisal. *International Journal of Contemporary Hospitality Management, 31*(4), 1819–1835. https://doi.org/10.1108/ijchm-11-2017-0737

Huppert, F. A. (2009). Psychological well-being: Evidence regarding its causes and consequences. *Applied Psychology: Health and Well-Being, 1*(2), 137–164. https://doi.org/10.1111/j.1758-0854.2009.01008.x

Hyun, K. K., & Yoan, R. S. (2018). Variations in the Asian collectivistic working culture in intercultural collaboration: A case of a South Korean company in Malaysia. *Pertanika Journal of Social Sciences and Humanities, 26*(3), 1633–1650.

Insurance Asia. (2023, September 21). *AIA Malaysia offers groundbreaking end-to-end mental health solution for employees.* Retrieved February 4, 2024, from https://insuranceasia.com/event-news/aia-malaysia-offers-groundbreaking-end-end-mental-health-solution-employees

Intellect. (2024, October 10). *ASEAN workplace wellbeing report 2024: Nearly half of Singapore employers are hesitant to hire staff with mental health conditions.* Retrieved April 11, 2025, from https://intellect.co/read/asean-private-sector-pr/

Jalil, M. F., & Ali, A. (2023). The influence of meaningful work on the mental health of SME employees in the COVID-19 era: Can coping strategies mediate the relationship? *BMC Public Health, 23*(1), 2435–2435. https://doi.org/10.1186/s12889-023-17347-3

Jalil, M. F., Tariq, B., & Ali, A. (2023). Does meaningful work mediate the relationship between empowering leadership and mental health? Evidence from Malaysian SME employees. *Frontiers in Sociology, 8*, 1138536. https://doi.org/10.3389/fsoc.2023.1138536

Johnson, A., Dey, S., Nguyen, H., Groth, M., Joyce, S., Tan, L., Glozier, N., & Harvey, S. B. (2020). A review and agenda for examining how technology-driven changes at work will impact workplace mental health and employee well-being. *Australian Journal of Management, 45*(3), 402–424. https://doi.org/10.1177/0312896220922292

Kaslow, N. J., Dunn, S. E., Bonsall, J. M., Mekonnen, E., Newsome, J., Pendley, A. M., Wierson, M., Schwartz, A. C., Sherman, M. D., Kaslow, N. J., & Graves, C. C. (2021). A Roadmap for patient- and family-centered care during the pandemic. *Couple and Family Psychology, 10*(3), 223–232. https://doi.org/10.1037/cfp0000176

Kelloway, E. K., & Drapeau, M. (2017). Mental health in the workplace: Towards evidence-based practice. *Canadian Psychology/psychologie Canadienne, 58*(1), 1–6. https://doi.org/10.1037/cap0000084

Kelloway, E. K., Dimoff, J. K., & Gilbert, S. (2023). Mental health in the workplace. *Annual Review of Organizational Psychology and Organizational Behavior, 10*, 363–387. https://doi.org/10.1146/annurev-orgpsych-120920-050527

Khor, L. K., & Tan, C. L. (2023). Workforce management in the post-pandemic era: Evidence from multinational companies using grounded theory. *Global Business and Organizational Excellence, 42*(4), 93–104. https://doi.org/10.1002/joe.22174

Kim, J., Chae, D., & Yoo, J. Y. (2021). Reasons behind generation Z nursing students' intentions to leave their profession: A cross-sectional study. *Inquiry (Chicago), 58*, 1–8. https://doi.org/10.1177/0046958021999928

Lee, Y. K. (2021). Impacts of digital technostress and digital technology self-efficacy on Fintech usage intention of Chinese Gen Z consumers. *Sustainability, 13*(9), 5077. https://doi.org/10.3390/su13095077

Lim, W. M. (2023). The workforce revolution: Reimagining work, workers, and workplaces for the future. *Global Business and Organizational Excellence, 42*(4), 5–10. https://doi.org/10.1002/joe.22218

Liswandi., & Muhammad, R. (2023). The association between work-life balance and employee mental health: A systemic review. *Asia Pacific Journal of Health Management, 18*(3), 263–280. https://doi.org/10.24083/apjhm.v18i3.2565

Lyles, S., Khan, Z., Qureshi, O., Shaikh, M., & De Luca, V. (2023). A systematic review of the effectiveness and implementation readiness of psychosocial interventions for psychosis in South Asia. *PloS One, 18*(11), Article e0283411. https://doi.org/10.1371/journal.pone.0283411

Mahikul, W., Srichan, P., Narkkul, U., Soontarawirat, I., Kitro, A., Pumipuntu, N., Saita, S., & Narayam, P. (2024). Mental health status and quality of life among Thai people after the COVID-19 outbreak: A cross-sectional study. *Scientific Reports, 14*(1), 25896–25922. https://doi.org/10.1038/s41598-024-77077-3

Maravilla, N. M. A. T., & Tan, M. J. T. (2021). Philippine Mental Health Act: Just an act? A call to look into the bi-directionality of mental health and economy. *Frontiers in Psychology, 12*, 706483–706483. https://doi.org/10.3389/fpsyg.2021.706483

Naji, G. M. A., Isha, A. S. N., Alazzani, A., Brough, P., Saleem, M. S., Mohyaldinn, M. E., & Alzoraiki, M. (2022). Do leadership, organizational communication, and work environment impact employees' psychosocial hazards in the oil and gas industry? *International Journal of Environmental Research and Public Health, 19*(8), 4432. https://doi.org/10.3390/ijerph19084432

Nnambooze, B. E., & Parumasur, S. B. (2016). Understanding the multigenerational workforce: Are the generations significantly different or similar? *Corporate Ownership and Control, 13*(2), 244–257. https://doi.org/10.22495/cocv13i2c1p4

Oakman, J., Kinsman, N., Stuckey, R., Graham, M., & Weale, V. (2020). A rapid review of mental and physical health effects of working at home: How do we optimise health? *BMC Public Health, 20*(1), 1825. https://doi.org/10.1186/s12889-020-09875-z

Panbarasan, M., Karthikeshwaran, R., Sudharsan, J., & Thamizhmani, V. (2020). Psychological ailments in the working environment of the oil & gas sector—An area of eminence in health risk globally. *International Journal of Scientific and Technology Research, 9*(1), 2987–2993. https://www.ijstr.org/final-print/jan2020/Psychological-Ailments-In-The-Working-Environment-Of-Oil-Gas-Sector-An-Area-Of-Eminence-In-Health-Risk-Globally.pdf

PETRONAS. (2022). *PETRONAS partners with Malaysian Psychiatric Association to safeguard employees' mental wellbeing.* Retrieved February 4, 2024, from https://www.petronas.com/media/media-releases/petronas-partners-malaysian-psychiatric-association-safeguard-employees-mental#:~:text=Under%20the%20partnership%2C%20MPA%20will,first%20aiders%20through%20the%20programme

QI Group. (2023, August 8). *QI Group's Malaysian HR team chalks up inaugural win at the Employee Experience Award 2023.* Retrieved February 4, 2024, from https://www.qigroup.com/news/qi-groups-malaysian-hr-team-chalks-up-inaugural-win-at-the-employee-experience-award-2023

Sarpong, S. (2024). Work-life balance: The challenges and the search for equilibrium in Malaysia. In D. Crowther, & S. Seifi (Eds.), *The Routledge companion to the future of management research* (1st ed., pp. 172–187). Routledge. https://doi.org/10.4324/9781003225508-13

Sweetland, A. C., Oquendo, M. A., Sidat, M., Santos, P. F., Vermund, S. H., Duarte, C. S., Arbuckle, M., & Wainberg, M. L. (2014). Closing the mental health gap in low-income settings by building research capacity: Perspectives from Mozambique. *Annals of Global Health, 80*(2), 126–133. https://doi.org/10.1016/j.aogh.2014.04.014

Tan, C. H., Koo, A. C., Rahmat, H., Siew, W. F., Cheang, A. W. O., & Amir Sharji, E. (2022). A quantitative study exploring the acceptance of the eHealth model for mental wellness among digital workers [version 1; peer review: 1 approved, 1 approved with reservations]. *F1000 Research, 11*, 111. https://doi.org/10.12688/f1000research.73482.1

Tan, K.-A., Keng, S.-L., & Abu Talib, M. (2023). Editorial: Mental health issues in Southeast Asia regions: Looking back and moving forward. *Frontiers in Psychiatry, 14*, 1229079–1229079. https://doi.org/10.3389/fpsyt.2023.1229079

The Star. (2022, August 16). *Mental health protection for the workforce*. Retrieved February 4, 2024, from https://www.thestar.com.my/news/nation/2022/08/16/mental-health-protection-for-the-workforce

van der Walt, F., & Lezar, L. W. P. (2019). Flourishing and thriving for well-being. In M. Coetzee (Ed.), *Thriving in digital workspaces* (pp. 85–107). Springer International Publishing AG. https://doi.org/10.1007/978-3-030-24463-7_5

Vadlamani, L. N., & Gowda, M. (2019). Practical implications of Mental Healthcare Act 2017: Suicide and suicide attempt. *Indian Journal of Psychiatry, 61*(Suppl 4), S750–S755. https://doi.org/10.4103/psychiatry.IndianJPsychiatry_116_19

Villarreal-Zegarra, D., Reátegui-Rivera, C. M., Sabastizagal-Vela, I., Burgos-Flores, M. A., Cama-Ttito, N. A., Rosales-Rimache, J., & Useche, S. A. (2022). Policies on mental health in the workplace during the COVID-19 pandemic: A scoping review. *PloS One, 17*(7), Article e0272296. https://doi.org/10.1371/journal.pone.027229

Wang, Y., Yan, L., & Alshuaibi, M. S. I. (2025). The great wall against burnout: CSR's influence in China's hospitality services sector. *Acta Psychologica, 255*, 104922. https://doi.org/10.1016/j.actpsy.2025.104922

Wells, J., Scheibein, F., Pais, L., Rebelo dos Santos, N., Dalluege, C.-A., Czakert, J. P., & Berger, R. (2023). A systematic review of the impact of remote working referenced to the concept of work-life flow on physical and psychological health. *Workplace Health & Safety, 71*(11), 507–521. https://doi.org/10.1177/21650799231176397

World Health Organization. (2005, January 1). *Promoting mental health: Concepts, emerging evidence, practice*. Retrieved February 27, 2024, from https://www.who.int/publications/i/item/9241562943

World Health Organization. (2022, June 16). *World mental health report: Transforming mental health for all*. World Health Organization. Retrieved February 27, 2024, from https://www.who.int/publications/i/item/9789240049338

Yamamoto, A., Nagata, T., Odagami, K., Adi, N. P., Nagata, M., & Mori, K. (2024). Relationship between companies' responses to near-miss reports and turnover intentions of workers: A nationwide cross-sectional study. *Safety and Health at Work, 15*(2), 187–191. https://doi.org/10.1016/j.shaw.2024.01.004

Yang, Q., Huo, J., & Jiang, Y. (2024). How to reduce the influence of COVID-19 epidemic on employees' anxiety of continuous work in China? Empirical analysis based on industrial enterprises. *Social Work in Public Health, 39*(8), 837–853. https://doi.org/10.1080/19371918.2024.2395263

Yap, C. Y., Chua, H. K., & Chin, Y. W. (2023). Into IR4.0: Charting the social and occupational changes in industrialised Malaysia. *Kajian Malaysia, 41*(2), 81–102. https://doi.org/10.21315/km2023.41.2.5

Yap, W. M., & Zainal Badri, S. K. (2020). What makes millennials happy in their workplace? *Asian Academy of Management Journal, 25*(1), 103–121. https://doi.org/10.21315/aamj2020.25.1.6

Yee, W. M., Al Mamun, A., Xueyun, Z., Hussain, W. M. H. W., & Yang, Q. (2023). Modelling the significance of psychological, social, and situational factors on work efficiency and the preference for working from home in Southeast Asia. *Heliyon, 9*(6), Article e17561. https://doi.org/10.1016/j.heliyon.2023.e17561

Yukawa, K., Shimanuki, M., & Eguchi, H. (2024). Prevention and future issues of karoshi and suicide by overwork in Japan. *Journal of the National Institute of Public Health, 73*(1), 16–31. https://doi.org/10.20683/jniph.73.1_16

Yusof, M. Z., Razali, M. A. E., & Mohd Rawi, M. F. A. (2022). Mental health status and psychosocial job determinants among employees in a Malaysian higher institution during the COVID-19 pandemic. *Malaysian Journal of Public Health Medicine, 22*(2), 68–76. https://doi.org/10.37268/mjphm/vol.22/no.2/art.1362

Zainal Badri, S. K., Mun, C. T. Y., & Ramos, H. M. (2020). Work passion and psychological wellbeing among millennial employees in Malaysia: The moderating role of personality. *International Journal of Employment Studies, 28*(1), 28–47.

Zarin, N. I., Kaur, K., Ismail, S., Yap, V. C., & Siong, C. S. (2022). A call for a new stress index for Malaysian teachers: A systematic review of studies of stress among teachers in Malaysia. *World Journal of English Language, 12*(2), 202–214. https://doi.org/10.5430/wjel.v12n2p202

Kiat Sing Heng joined the School of Business, Faculty of Business, Design and Arts, Swinburne University of Technology Sarawak Campus as a Lecturer in 2007. She served in various academic management and leadership roles, including Associate Dean (Accreditation and Curriculum Enhancement) and Program Coordinator for Undergraduate Business: Non-accounting and Postgraduate Business courses. She teaches economics subjects for diploma and undergraduate business courses and organisational behaviour for undergraduate and postgraduate business courses. She was also an internal auditor, Malaysia Research Assessment (MyRA) Internal Audit, appointed by the Ministry of Higher Education Malaysia from November 2017 to October 2020.

Chia Hua Sim is an academic staff member at the Faculty of Business, Design, and Arts, Swinburne University of Technology Sarawak Campus, with over ten years of experience in academia. She currently facilitates staff participation in professional development activities, including supporting the development and enculturation of new faculty members. Collaborating with Digital Learning Technologies at the Sarawak Campus and the Melbourne counterpart, she promotes new educational initiatives. Additionally, she works closely with faculty representatives to identify and prioritize skill enhancement areas, fostering the continuous academic and professional development of faculty members.

Mei Yuan Law is a digital technologist and lecturer at the School of Design and Arts, Faculty of Business, Design, and Arts, Swinburne University of Technology Sarawak Campus. Before joining Swinburne Sarawak as a full-time academic, she worked in the architecture industry and freelanced as an online verified streamer for a start-up blockchain company while serving as a sessional lecturer for over five years. She has published research on higher education and actively contributes to non-traditional research outputs (NTRO). She is a reviewer for several indexed journals and is pursuing a Doctorate in Education at Open University Malaysia, focusing on instructional technology.

Chapter 6
Reimagining Women's Future Workplaces: Where Are We Heading?

Clusters of Future Studies: Corporate Foresight; Past and Futures; Humanity at the Limen; Technological Trends

Donna Barclay, Nivakan Sritharan, and Jacob King Soon Ting

Abstract This chapter is a conceptual paper that explores the evolving role of women in the workforce, focusing on progress, persistent challenges, and solutions, particularly for working mothers. The chapter aims to evaluate systemic barriers such as cultural norms, discriminatory practices, and the glass ceiling while assessing the effectiveness of flexible work arrangements (FWA) and technology-driven work models in promoting gender equity. The chapter explores research and case studies from Asia and other regions to identify trends, challenges, and opportunities. It highlights how the COVID-19 pandemic accelerated the adoption of technology-driven work models, such as virtual collaboration tools and ICT, fostering flexibility and inclusivity. However, findings reveal persistent gender disparities in access to flexible work arrangements, with men often benefiting more from existing policies, especially in Asia. Key takeaways include the need for organisations to regularly update policies to promote diversity, address employee needs, and create equitable workplaces. The chapter emphasises the importance of collaboration between governments and businesses in establishing family-friendly policies and legal standards. It also stresses the critical role of technology in advancing flexible work practices, supporting re-entry programmes for female workers, and fostering inclusive, future-ready workplaces.

Keywords Glass ceiling · Gender equity · Flexible work arrangements (fwa) · Technology-driven work arrangements · Work-life balance

D. Barclay (✉) · N. Sritharan · J. K. S. Ting
Swinburne University of Technology Sarawak Campus, Kuching, Malaysia
e-mail: dtan@swinburne.edu.my

N. Sritharan
e-mail: nsritharan@swinburne.edu.my

J. K. S. Ting
e-mail: jting@swinburne.edu.my

C. Y. Fung et al. (eds.), *Future of Work in Asia*,
https://doi.org/10.1007/978-3-031-95804-5_6

6.1 Introduction

The future of women's workplaces is at a pivotal juncture, shaped by the interplay of historical struggles, technological advancements, and cultural transformations. Over the past few decades, women have significantly increased their presence across diverse industries, becoming vital contributors to the global workforce (Atan et al., 2021; Oo & Lim, 2021). Yet, despite this progress, systemic barriers continue to hinder their career advancement, particularly for working mothers. These obstacles, including unequal domestic burdens, limited childcare options, and rigid workplace structures, emphasise the urgent need to reimagine workspaces that are inclusive and equitable (Kaur, 2004; Yadav & Jadhav-Tilekar, 2021).

This chapter critically examines how evolving work arrangements, particularly those driven by technology, are shaping opportunities for women in the workforce. A systematic literature review (SLR) was conducted to synthesise findings on remote work, flexible arrangements, and their implications for work-life balance and gender equality. Drawing on high-quality studies, this analysis highlights the potential of flexible work arrangements to empower women while exposing persistent cultural and structural barriers to their widespread adoption (Mushfiqur et al., 2018; Thilagavathy & Geetha, 2021).

In Asia, where cultural norms and economic disparities play a significant role, the challenges faced by women are multifaceted. Historically, women in the region have been constrained by patriarchal structures, even though pre-colonial Southeast Asia saw relatively high levels of economic autonomy for women (Booth, 2016; Reid, 1988). While modernisation and improved education have facilitated greater workforce participation, unequal access to decent work opportunities and discriminatory workplace policies persist (Baird et al., 2023; International Labour Organization, 2023).

Structural issues, such as the invisible artificial barrier that hinders women's career progress (the 'glass ceiling'): Watanabe and Kwarteng (2024), gender biases, and systemic discrimination, further exacerbate inequalities. These barriers are particularly evident in male-dominated industries, where inflexible schedules, gender stereotypes, and limited career progression opportunities prevail (Desai & D'souza, 2024). Achieving equitable workplaces for women requires not only organisational reforms but also legal, institutional, and cultural shifts to dismantle entrenched obstacles (Jain, 2023; Williamson & Colley, 2023).

This chapter envisions a future where women's workplaces are reimagined to foster inclusivity, balance, and empowerment. By addressing systemic challenges and leveraging the potential of FWA, societies can unlock the full potential of women in the workforce, paving the way for meaningful gender equality.

6.2 Key Concepts and Issues

6.2.1 *Traditional and Contemporary Work Arrangements for Women in Asia*

Women's participation in the workforce in Asia has undergone significant transformation (refer to Fig. 6.1), influenced by cultural, economic, and political shifts, over the past decades. Historically, women's roles were largely confined to domestic duties and agricultural work, often under rigid patriarchal structures that limited their economic autonomy. However, industrialisation in the 19th and early twentieth centuries began to change this dynamic, drawing women into factories and other employment in cities. Following World War II and the economic booms in countries like Japan, South Korea, and China, female workers entered diverse sectors, including manufacturing, services, and education (Boserup et al., 2007).

By contrast, a comprehensive study of Southeast Asia in the pre-colonial era suggests that women enjoyed both economic and social autonomy as compared with women in China, India and the Arab world. In fact, it was not unusual for pre-colonial Southeast Asian women to work outside the home in textile and ceramic production; and they participated not only in the bazaar economy but also in large-scale trade and other industries (Reid, 1988). According to Booth (2016), the trend of women

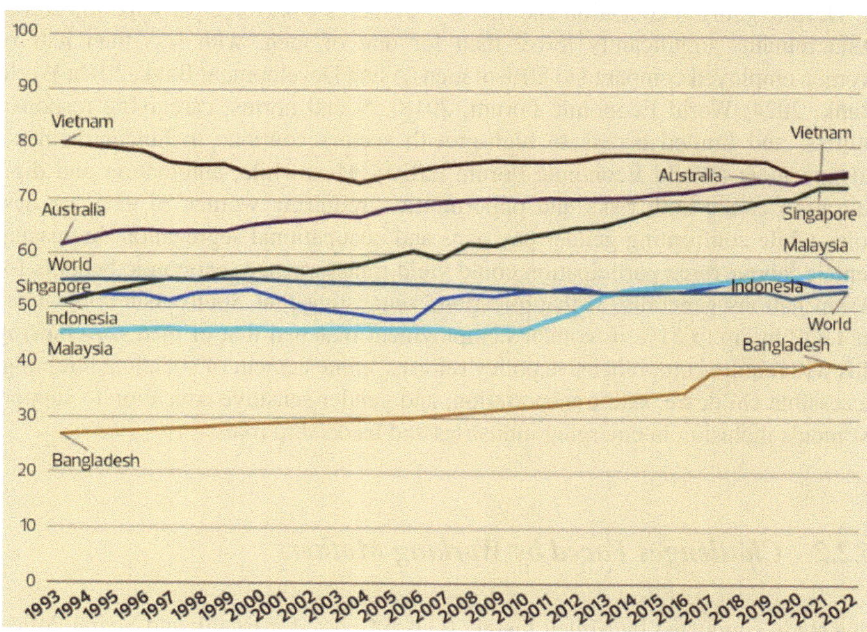

Fig. 6.1 Trends in female labour force participation in Australia and Asia from 1993 to 2022. *Source* The Edge Malaysia (4th April, 2024)

participating in the labour force continued after colonisation albeit in a different pattern.

By the 1960s, most countries in Asia had thrown off the shackles of colonialism. Coupled with an increase in female literacy, this meant that women participated in employment in both rural and urban areas to a reasonably large extent, even accounting for 50% of the labour force in countries such as Burma (as it was then known), Thailand, Cambodia, Laos, the Philippines and Vietnam. By the twenty-first century, modernisation and better educational opportunities for women led to an increasing number of women entering the labour market. However, the evolving work environment presented significant challenges for female employees in balancing work commitments and family responsibilities, an issue that resonates to this day (Yadav & Jadhav-Tilekar, 2021).

A joint report by the International Labour Organization and Asian Development Bank (2023) examines sectoral employment trends of both males and females in Asia, analysing work patterns through the lens of gender equality, inclusive growth, and social justice. Women remain heavily concentrated in low value-added sectors such as agriculture and retail trade, where decent work opportunities are scarce. Over the past thirty years, sectors experiencing the most substantial job growth in Asia have consistently favoured men, especially in rapidly expanding fields like information technology, where women have secured only a quarter of available jobs. In contrast, the only sector where women saw a significant share of job growth was accommodation and food services, with over half of new jobs going to women.

Despite gains in education and literacy, women's workforce participation across Asia remains significantly lower than for that of men, with less than half of women employed compared to 80% of men (Asian Development Bank, 2016; World Bank, 2024; World Economic Forum, 2018). Social norms, caregiving responsibilities, and limited access to high-growth sectors continue to hinder women's advancement (World Economic Forum, 2023). Meanwhile, automation and digitalisation create both risks and opportunities, requiring women to navigate new roles while confronting gender pay gaps and occupational segregation. Increasing female labour force participation could yield transformative economic benefits for Asian nations generally, with projections suggesting that South Asia could raise its GDP by up to 51% if women's employment matched that of men's. Achieving this will require comprehensive policy reforms, enhancement of vocational training, accessible childcare, safe transportation, and gender-sensitive education to support women's inclusion in emerging industries and leadership roles.

6.2.2 Challenges Faced by Working Mothers

In a study conducted by Milieu Insight (Galen Centre for Health and Social Policy, 2024), six in ten working mothers in Southeast Asia reported struggling to balance family commitments and work responsibilities. Specific concerns included finding reliable childcare options and a lack of supportive workplace policies. Interestingly, women also reported feeling guilty when they prioritise work over family. The Galen research is reinforced by the ILO (2023) study, which reports that Asian women encounter structural barriers that hinder their full participation in the labour market. A significant issue common across all parts of Asia is the unequal burden of unpaid care or domestic duties borne by women. Thus, it is clear that one of the primary reasons that married female employees choose to exit the workforce is the lack of workplace flexibility and the difficulty in maintaining a healthy work-life balance (Kaur, 2004).

In *Women and Work in Asia and the Pacific: Experiences, Challenges and Ways Forward*' the authors explore the diverse experiences of women in the workforce across Asia and the Pacific. They emphasise the obstacles women encounter in entering and succeeding in employment, analyse multiple factors affecting their participation, and propose strategies to promote gender equality in workplaces. The authors explore cultural, social, economic, and policy factors influencing women's employment experiences and offer suggestions to enhance inclusivity and empower women in the labour market. (Baird et al., 2023).

The challenges faced by working mothers in Southeast Asia have significant implications for the future workforce (Gideon, 2024; Ha, 2024). The persistent struggle to balance work and family responsibilities, compounded by inadequate childcare options, workplace inflexibility, and the unequal burden of domestic duties, often leads to career interruptions or exits from the workforce (HR Asia, 2024; Pfordten, 2024). This limits women's economic contributions and hinders gender equality in the labour market. The guilt associated with prioritising work over family and the lack of supportive policies, such as extended maternity leave and FWA (Ho et al., 2024). Addressing these challenges through inclusive workplace policies, remote work options, and better support systems can enhance women's participation in the workforce, improve their well-being, and unlock substantial economic potential for the region.

6.2.3 Current Trends in Flexible Work Arrangements (FWA)

Research has been conducted globally to assess the availability and demand for FWA. However, as noted by Thilagavathy and Geetha (2021), most published studies on work-life balance are concentrated in the United States, Europe, and Asia. By contrast, there is still a lack of research in other regions, including parts of Asia, and this gap persists today.

Masuda et al. (2012) conducted the first empirical study examining differences in the availability of FWA across Latin American, Anglo, and Asian countries. The study revealed that workers in individualistic, Anglo countries were more likely to have access to flexitime, compressed workweek models, and telecommuting options, whereas collectivistic countries in Asia and Latin America lag behind. This highlights the influence of cultural norms on FWA adoption. Similarly, in Africa, patriarchal and

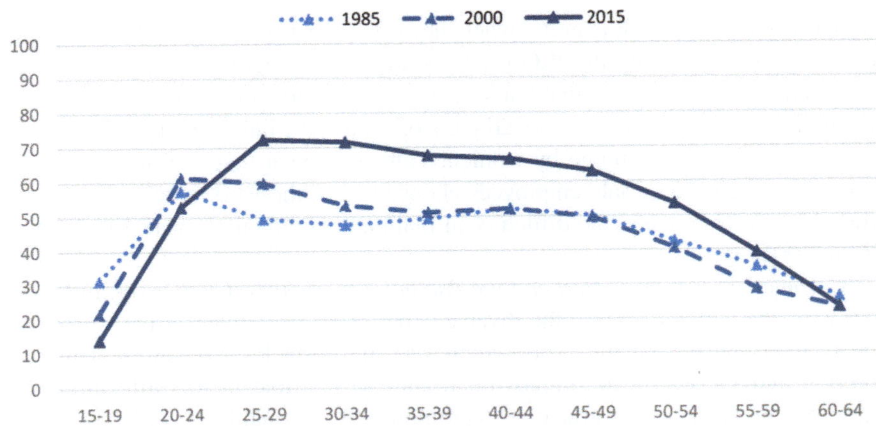

Fig. 6.2 Female labour force participation rate by age cohort in Malaysia. *Source* Syed Salleh and Mansor (2022, p. 3)

cultural norms often hinder women's efforts to achieve work-life balance (Mushfiqur et al., 2018).

Researchers in Malaysia have found different factors at play. Age is one of the factors shows significant participation change among female employees (refer to Fig. 6.2). Malaysia's economic transition from an agricultural to an industrial economy in the 1970s led to a significant increase in employment opportunities for women. This shift resulted in two key trends: a rise in dual-income households and an increase in the number of female university students, surpassing their male counterparts. However, the challenge of balancing work and family responsibilities has contributed to persistently low female labour force participation (Subramaniam et al., 2015). FWA could provide a viable solution, enabling women to better manage childcare and eldercare responsibilities while enhancing their focus and productivity at work (Parker, 2019).

The future of work will undoubtedly encompass a significant recognition of the role of flexible work arrangements (FWA). Seo and Kim (2024) contend that organisations must explore a diverse range of FWA, which can be broadly classified into flexitime and flexplace. Flexitime refers to flexible working hours, including adjusted start and end times and compressed workweeks. Flexplace, on the other hand, enables employees to work from remote locations, including home. Such policies accommodate individual scheduling and location preferences, thereby enhancing both employee well-being and organisational outcomes by enabling employees to allocate time and energy more effectively across personal and professional responsibilities. Additionally, FWA promote collaboration across various locations, fostering greater adaptability in the modern workforce (Shifrin & Michel, 2021).

FWA is one tool in the management toolbox that may address labour shortages, given the phenomena of ageing populations in many Asian nations. In their 2018, pre-pandemic study, Loichinger and Cheng found that Hong Kong, Japan, South Korea, and Singapore were already facing labour shortages. This study analysed past trends and projected future labour supply up to 2050. Better-educated women can reduce labour shrinkage. This stresses the importance of education when considering future labour supply. Additionally, in 2019 McKinsey predicted that by 2025, women in Asia could contribute $4.5 trillion to GDP. India and China have the biggest potential in this regard (18% and 13% respectively). However, Japan, Singapore and mature Asian economies also stand to benefit from attracting talented women into workplaces. FWA shows a great deal of promise in achieving this end, however, companies that offer remote or hybrid work need to implement comprehensive policies covering all aspects of working from home (Javed, 2024).

6.2.4 Technological Advancements and Their Impacts

The role of technology in enabling remote work and flexible schedules has been extensively documented in various studies. These studies emphasise the impact of technology on work-life balance, productivity, and job satisfaction. For example, virtual collaboration tools and remote work platforms allow employees to manage flexible schedules and work remotely, which helps address mental health challenges and promotes work-life balance (Peslak et al., 2023). Similarly, the adoption of information and communication technology (ICT) during the COVID-19 pandemic has facilitated the widespread use of remote practices in various activities, such as work and virtual tasks. This has enhanced attitudes towards remote work, perceived benefits, and trust (Kwek et al., 2024). In addition, research by Kaakandikar and Gawande (2023) highlights the role of digital infrastructure in transforming traditional office settings permitting seamless collaboration and promoting diversity and inclusion within virtual teams. Figure 6.3 shows the milestones in the emergence of the move towards flexible work models, with its origins in telecommuting in 1972 to the present reliance of various types of various FWA models.

Globally, women continue to perform a disproportionate share of unpaid labour as compared to men (see Fig. 6.4). Advances in technology have facilitated FWA for women, enabling them to navigate the dual responsibilities of professional employment and domestic labour (Dhanya & Kinslin, 2017). For women engaged in digital entrepreneurship, technological advancements provide essential tools and platforms for remote business management, fostering innovation and enhancing business satisfaction (Chakraborty et al., 2019). However, while technology has the potential to enhance work flexibility, it can also reinforce traditional gender roles and perpetuate existing inequalities. Parry (2024) highlights this disparity, noting that fathers benefit more significantly from spatial and temporal work prioritisation than mothers. Similarly, Rodríguez-Modroño and López-Igual (2021) observe that while home-based teleworking can improve the quality of working time for women, it often comes at the cost of reduced income and limited career advancement opportunities. These findings reinforce the complex relationship between technology, gender dynamics, and labour market outcomes.

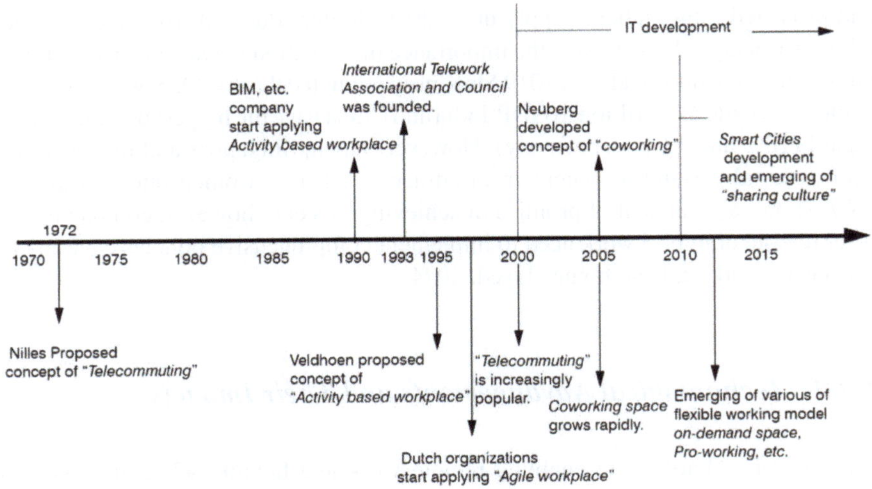

Fig. 6.3 History of the flexible work model from 1970. *Source* Yu et al., (2019, p. 450)

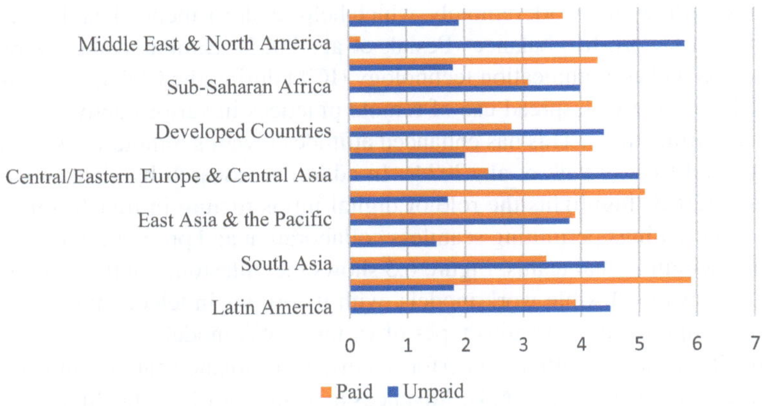

Fig. 6.4 Average total hours of paid and unpaid work done per day in regions of the world. *Source* Armstrong (8th March 2019)

Apart from this, the importance of technology in overcoming geographical and institutional barriers is also evident. Technology empowers women entrepreneurs in regions like Jaipur, Rajasthan, India by facilitating remote business operations, enabling them to overcome geographical constraints and manage businesses flexibly (Shastri et al., 2019). Technology also plays a pivotal role in enabling remote work for women, supporting flexible work timings that address work-related stress and work-life imbalance (Fathima et al., 2020). Moreover, technology supports remote work for software professionals in emerging economies like India, addressing ethical concerns and enhancing virtual work environments (Gupta & Pathak, 2024). All these studies illustrate how technology not only enable flexible schedules and remote work but also challenge traditional norms and support diverse work arrangements.

6.2.5 *Tools and AI in Managing Work-Life Balance*

The COVID-19 pandemic created an urgent need to pivot to remote work and FWA to maintain workforce productivity and well-being (Parry, 2024). However, gendered patterns emerged, with mothers experiencing more fragmented work, highlighting persistent inequalities in work-life balance. More broadly, studies on work-life balance emphasise common themes such as flexibility, technological advancements, and supportive work environments. Peslak et al. (2023) highlight the role of FWA and comprehensive mental health programmes in fostering inclusive workplaces and supporting the psychological well-being of IT developers. Similarly, Pillai and Prasad (2023) identify training initiatives, human resource interventions, and coworking spaces as critical mechanisms for enhancing flexibility, improving productivity, and mitigating health risks in remote and hybrid work models. These findings collectively reinforce the evolving nature of work and the structural supports necessary to ensure equitable outcomes.

Technological advancements play a significant role in facilitating work-life balance. In their study, Kaakandikar and Gawande (2023) discuss how advances in technology and digital infrastructure are important for improving work-life integration, productivity, and efficiency. They also articulate how these advancements expand talent pools and reduce administrative costs. Similarly, Kwek et al. (2024) emphasise the use of information and communication technology (ICT) and social media for virtual events, which enable remote work and virtual classes. Ganguly et al. (2022) examine the advantages of virtual workspaces, hybrid work models, and remote work arrangements with geographical requirements for white-collar professionals. While the study does not specifically address gender, its findings highlight the potential of these work structures to mitigate challenges such as social isolation, technological difficulties, and workplace monitoring. These insights may hold broader applicability for similar employment contexts globally.

Supportive work environments and policies are essential in facilitating work-life balance, particularly for women. Fathima et al. (2020) report the importance of FWA, management policies that actively support women, equitable performance evaluations, and mentorship programmes in reducing work-related stress and promoting work-life balance. Similarly, Chakraborty et al. (2019) identifies non-monetary factors such as material comfort, and job satisfaction as key predictors of well-being and satisfaction among women in digital entrepreneurship. Research by Rodríguez-Modroño and López-Igual (2021) explores the relationship between telework arrangements and job quality, revealing that occasional teleworkers tend to report the most favourable work-life balance outcomes. Arora and Kumari (2022) emphasise the benefits of FWA and hybrid work models, which provide remote work opportunities and support the reintegration of women who have taken career breaks. Naaman's (2022) article suggests the advantages of teleworking, particularly for women and parents of young children. These studies collectively suggest that work arrangements tailored to the needs of women can significantly enhance work-life balance, offering broader implications for promoting gender equity in the workplace.

6.2.6 Potential Benefits and Drawbacks of Technology-Driven Work Arrangements

Technology-driven work arrangements offer numerous benefits, particularly in terms of flexibility, productivity, and work-life balance. Studies have shown that remote work and FWA can significantly improve employee well-being and satisfaction by allowing for better integration of work and personal life and reducing commuting time (Naaman, 2022; Rodríguez-Modroño & López-Igual, 2021). These arrangements are especially beneficial for women, as they provide the flexibility needed to juggle professional and personal responsibilities, promote gender inclusivity, and support mental health through professional development opportunities (Parry, 2024; Peslak et al., 2023). Moreover, the adoption of virtual workspaces, hybrid work models, and co-working spaces is considered the future of work, offering substantial time, cost, and effort savings, as well as access to diverse and global talent pools (Ganguly et al., 2022; Kaakandikar & Gawande, 2023; Kwek et al., 2024).

However, the drawbacks of technology-driven work arrangements cannot be overlooked. Challenges such as social isolation, technological difficulties, and ineffective communication often arise, leading to increased health risks, burnout, and job attrition, especially during extended periods of remote work (Dhanya & Kinslin, 2017; Pillai and Prasad, 2023). For highly mobile teleworkers, job quality and work-life balance suffer due to the blurring of boundaries between work and home life, reinforcing traditional gender norms and contributing to gender-based challenges such as discrimination (Peslak et al., 2023; Rodríguez-Modroño & López-Igual, 2021). Moreover, over-reliance on technology can exacerbate feelings of isolation and hinder team cohesion and conflict management in virtual settings (Arora & Kumari, 2022; Kaakandikar & Gawande, 2023).

Furthermore, ethical considerations and societal challenges make technology-driven work arrangements even more complex. Addressing ethical dilemmas related to people, technology, and culture is crucial, and implementing comprehensive measures to enhance virtual work experiences (Gupta & Pathak, 2024). While female scientists and entrepreneurs benefit from FWA and increased innovation opportunities, they still face significant obstacles, such as work-related stress, sexual harassment, and a lack of recognition for their professional capabilities (Fathima et al., 2020; Shastri et al., 2019). These studies highlight the importance of creating inclusive work environments and implementing robust support systems to promote the mental health and well-being of all employees involved in technology-driven workplaces.

6.2.7 *Government and Policy Influences*

Women bring a fresh and diverse perspective to businesses in competitive markets. Research indicates that investing in women leads to significant social and economic benefits. The empowerment of women through flexible mechanisms related to employment brings positive outcomes to businesses (Taneja et al., 2012). In 2022, Deloitte's Women @ Work Report, which surveyed 500 women as part of a larger global study, revealed significant issues related to work flexibility, mental health, and workplace inclusion. Only 33% of women reported that their workplace offers FWA, while over 94% stated that requesting flexible work would affect their likelihood of promotion.

6.2.8 *Government and Policy Responses by Asian Nations*

With Deloitte (2022) predicting a continuation of the 'Great Resignation' in the post-pandemic landscape, government intervention via laws and regulations that offer robust support to women with care obligations is one way forward to ensure a sufficient labour pool. Additionally, the accelerating pace of digital transformation will be pivotal in reshaping global and regional work arrangements, particularly with the rise of hybrid and remote work models. Technology's role will extend beyond enabling new ways of working; it will influence workforce inclusivity by providing flexible work options for marginalised groups, including working mothers.

Legislative and policy efforts across Asia reflect a growing recognition of the importance of FWA in supporting workforce diversity and productivity, albeit with varying degrees of implementation and support infrastructure. In order to fully support women, especially those with children, further developments in the uptake of flexible work patterns will be crucial in shaping future workplace norms and, in turn, in supporting economic growth across the region.

(a) **Malaysia**

One Asian nation with legislation that supports the needs of women with care obligations who desire to work is Malaysia. Part XIIc of the *Employment Act 1995* (EA 1955), entitled 'Flexible Working Arrangement,' permits employees to apply for adjustments to their work patterns. The amendment came into force in 2023 and provides that employees may make a written application to their employer seeking flexible work (s. 60Q(1)). The employer has sixty days to reply to the request, this response must be in writing, and the employer must provide reasons for any refusal (s. 60Q(2)). It is expected that this option for flexible work in the Malaysian context will encourage both the hiring and retention of female employees. A recent paper by Teoh et al. (2024) investigates the readiness of female employees in Malaysia for FWA. The study identifies key factors influencing this readiness to adopt FWA, including FWA benefits, career advancement, and company policy, while family responsibilities and supportive workplaces have less impact (Teoh et al., 2024).

(b) **Singapore**

Following the lead of other developed nations, Singapore, introduced the right of employees to request FWA, on 1 December 2024, albeit via mandatory guidelines rather than through legislation. As with the law in Malaysia, employers in Singapore will be required to respond to any such request in writing and to provide cogent reasons should the request be denied. Ms Gan Siow Huang, Minister of State for Manpower, and co-chair of the Tripartite Workgroup that proposed the FWA scheme stated that once implemented, the requests for FWA "will shape norms and expectations around FWA so that our workplaces can be more inclusive and productive" (Ministry of Manpower, 2024). As of the time of writing, it remains premature to fully assess the impact of these new mandatory guidelines.

(c) **Indonesia**

The Manpower Law of 2003 establishes a comprehensive framework for labour regulations, including employment terms, working conditions, and labour rights. However, while Minister of Manpower Regulation No. 11 of 2019 acknowledges flexibility in work arrangements, it does not explicitly grant employees the right to request flexible work arrangements (FWA). One major obstacle in Indonesia's shift toward FWA is the limited internet infrastructure, which hinders the widespread adoption of flexible work practices, especially in rural areas. Women's economic participation thus remains low due to societal norms positioning mothers as primary caregivers and the absence of FWA in the formal sector, contributing to gender inequality (Cameron, 2023).

(d) **China**

In 2021, China announced a policy allowing married couples to have up to three children to optimise the nation's demographic structure. This policy shift, as outlined on the State Council's website, aims to address the country's aging population and declining birth rates. To support this initiative, the government has committed to reducing the costs associated with childbearing, improving pre- and post-natal services, and enhancing nursery facilities. However, despite the significant number of infants and children—over 40 million under the age of three—only 5.5 percent of parents have access to nursery services, as reported by Yang Wenzhuang, Head of the Population and Family Department of the National Health Commission (State Council Information Office, 2021). This lack of adequate childcare services is a key factor contributing to women's exit from the workforce. At present, China does not have a specific policy granting employees the right to request FWA, further exacerbating the challenges faced by working parents, particularly women, in balancing work and family responsibilities.

(e) **Thailand**

While Thailand does not have specific laws mandating FWA like some other countries, there is a growing acceptance and adoption of flexible work practices among employers, reflecting broader trends towards modernising work arrangements and supporting workforce well-being. For example, s. 23/1 of the *Labour Protection Act* (Government of Thailand, 2023). While this section is not a wholesale implementation of flexible work rights, it does provide a guide that permits an employer and employee to enter into an agreement whereby the latter may work from home. The English translation of this section reads as follows:

> For the benefit of the employer's business operation and to promote quality of life and work performance of the employee, or in the case where it is necessary, the employer and the employee may agree that the employee can bring the work in the course of employment or that which is agreed with the employer — having the nature or description whereby the employee may perform such work conveniently outside the employer's establishment or premises — to perform such work at his/her home or residence, or agree that the employer may perform the work through use of information technology from elsewhere remotely (Nishimura & Asahi, 2023).

(f) **Japan**

The Parliament of Japan recently enacted a law mandating that companies offer flexible working options to employees with young children, aimed at helping them balance work and childcare responsibilities. This legislation includes updates to childcare and nursing care leave laws as part of a broader initiative to ensure that career development is not hindered by gender or family obligations. Key measures, to be implemented commencing in April 2025, require companies to provide at least two flexible working options (e.g., remote work, shorter hours) for employees with preschool-aged children. Additionally, efforts must be made to allow parents of children under three to work from home. The law extends the age for exemption from overtime work and the eligibility for childcare leave to accommodate school-related activities and illness care for children up to third grade (Japan Times, 2024).

6.3 Discussion of Implications

Future policy directions and potential implications on work arrangements in Asia will be influenced by a variety of factors, including technological advancements, economic shifts, and evolving societal expectations. For example, it is predicted that by 2030, the number of digital jobs globally will rise significantly, offering higher-paid salaries and addressing skill shortages. However, challenges like digital literacy and equitable access to technology remain critical issues that need addressing. For women in countries with good access to high-speed internet, there will be more opportunities than ever for work.

In Asia, countries like Singapore and Japan, with their advanced digital infrastructure and robust policy frameworks, are well-positioned to capitalise on this shift. However, disparities in digital access and literacy across less developed countries in the region, such as parts of Southeast Asia, could exacerbate existing inequalities. Governments will need to focus on reducing these gaps by investing in widespread internet access, digital skills training, and infrastructure development. This will allow a broader range of workers, particularly women, to participate in the digital economy, levelling the playing field across sectors and regions.

Apart from digital transformation, it is expected that hybrid work, which synchronises remote work with time spent in the office, will be leveraged by companies to balance flexibility and collaboration between co-workers. Many businesses across Asia have taken up this approach to meet changing employee expectations and improve work-life balance. The success of hybrid work depends on technological adaptation and is more suited to some Asian countries than others. Ahmed Mazhari, President of Microsoft Asia in 'The Future of (Hybrid) Work' podcast (2022) stated that "If done right, flexible work arrangements could be a very powerful tool for inclusion and for companies to benefit from a more diverse talent pool (McKinsey, 2022).

It must be emphasised that in order to promote equity, diversity and inclusion within workplaces, companies need to support gender equality and ensure that flexible work options are available to all employees, even in the absence of legislation. Governments in several Asian countries have enacted legislation to create workplace environments where career development is not stymied by family responsibilities. Legislative changes in Malaysia, Singapore and Japan are notable in this respect.

6.4 Future Directions

6.4.1 Creating Future-Ready Workplaces

To create truly inclusive workplaces, organisations must consistently evaluate and enhance their policies and practices. This involves regularly assessing employee needs, implementing diversity and inclusion training programs, and ensuring that leadership is dedicated to fostering an equitable environment. Furthermore, utilising technology to facilitate flexible working arrangements can further promote inclusivity. By prioritising these initiatives, companies can provide better support for working mothers, resulting in a more diverse, innovative, and productive workforce.

6.4.2 Integrating Technology

Looking ahead, the integration of technology into work arrangements is poised to significantly transform work-life balance, productivity, and job satisfaction. Emerging trends indicate an increasing reliance on virtual collaboration tools and remote work platforms, offering greater flexibility in schedules and enhancing opportunities for remote work. These advancements have the potential to mitigate mental health challenges and improve work-life balance. As digital infrastructure continues to evolve, traditional office environments will be reshaped, fostering seamless collaboration, boosting productivity, and promoting diversity and inclusion within virtual teams. However, it is crucial to address the gendered dimensions of remote work to ensure that technological innovations do not inadvertently reinforce traditional gender roles or perpetuate existing inequalities. Targeted interventions will be necessary to promote gender equity in this new landscape. Additionally, technology presents significant opportunities for empowering women entrepreneurs and professionals by overcoming geographical and institutional barriers, enabling remote business operations, flexible work hours, and reducing work-related stress and imbalance. As technology advances, it will contribute to more equitable and efficient work arrangements, challenging conventional norms and stimulating innovation across various sectors.

6.4.3 Future Work Practices

Future trends in managing work-life balance for working mothers will focus on using advanced AI technologies to improve flexibility and productivity. By integrating digital infrastructure and ICT, remote and hybrid work models will be facilitated, which will expand talent pools and reduce costs. It will be essential to establish comprehensive mental health programmes and create supportive work arrangements, including fair performance evaluations and mentorship programs, to promote inclusive workplaces. Technological advancements will also address social isolation and enhance HR practices, while the implementation of hybrid workplaces and flexible schedules will support women who are re-entering the workforce after taking career breaks. As technology continues to evolve, it will be crucial to ensure gender equity in these flexible arrangements to foster a balanced and inclusive work environment for women who wish to return to work after taking a career break.

6.4.4 Future Arrangements for Working Mothers

Future trends and directions for working mothers will increasingly rely on technology to enhance flexibility, productivity, and work-life balance. Organisations will need to prioritise the development and refinement of virtual workspaces, hybrid work models, and co-working spaces to support diverse talent pools and foster gender inclusivity. To address the challenges of social isolation and technological difficulties, there will need to be a focus on creating inclusive environments, comprehensive mental health programmes, and robust support systems that cater to women's needs. Ethical considerations and societal challenges, such as gender discrimination and the reinforcement of traditional gender roles, will be critical areas of focus. By leveraging technological advancements, implementing supportive policies, and fostering inclusive work cultures, future work arrangements will aim to provide equitable opportunities and promote the well-being of working mothers.

6.4.5 Imagining the Future of Work for Mothers

To create a truly inclusive workplace, organisations should take specific actions such as offering gender-neutral parental leave, flexible work hours, and dedicated childcare support. In some Asian countries like Malaysia, the government has already mandated paid family leave. This initiative should be considered at the national level in countries across Asia in conjunction with the establishment of minimum standards for employer-provided childcare services. Companies should also implement mandatory diversity and inclusion training for leadership and employees, ensuring accountability through regular audits of workplace equity. By leveraging technology to facilitate remote and hybrid work, employers can enable flexible schedules that better support working mothers while maintaining productivity and inclusivity.

Looking ahead, governments should also invest in improving digital infrastructure to enable reliable access to remote work for all employees, including those in rural areas. Industry leaders should adopt AI-driven project management tools and virtual collaboration platforms to support more flexible and efficient work environments. Policy frameworks must address the risk of exacerbating gender inequalities through remote work by establishing gender equity standards for performance evaluations, compensation, and promotions. This will prevent remote work from sidelining women or reinforcing traditional gender roles.

In addition to flexible arrangements, employers need to be cognisant of including mental health programmes, including free counselling services, stress management workshops, and mental health days to support working mothers. Governments can offer tax incentives for companies that adopt family-friendly policies, including extended parental leave and re-entry programmes for women returning from career breaks. Mentorship programs should be developed to ensure working mothers receive guidance and support in navigating career progression and maintaining work-life balance.

To support future work arrangements for working mothers, governments and organisations must collaborate to set legal standards for hybrid workplaces, ensuring inclusivity and fairness. Co-working spaces and community childcare centres should be established, particularly in underserved regions, to facilitate working mothers' access to professional opportunities. Ethical considerations must be addressed, with industries adopting policies that prevent discrimination and ensure equal access to technological resources. This combination of government policy, organisational action, and technology will foster a more equitable and inclusive workforce for working mothers.

6.5 Conclusion

Over the past few decades, there has been a notable increase in the number of women participating in the workforce, which has brought about fundamental changes in various industries. However, women still encounter multiple barriers that challenge their full participation and advancement. These barriers are influenced by structural and societal changes, and there are significant variations in their prevalence across different regions and industries. Additionally, intersectional factors such as race, class, and ethnicity further complicate women's experiences in the workforce, resulting in diverse challenges in different sectors. In male-dominated fields such as construction, women face ongoing obstacles, including long working hours, remote work locations, and inflexible schedules. The existence of the glass ceiling remains a major barrier, as issues like sexual harassment, pay disparities, and limited opportunities for career progression are exacerbated by societal expectations regarding women's domestic responsibilities.

Modernisation in the twenty-first century has created more opportunities for women in the job market, especially in regions like Asia and the Pacific. In Malaysia, for example, the shift from agriculture to industry in the 1970s resulted in more job openings for women. Women in the region still encounter obstacles in accessing quality employment opportunities. The COVID-19 pandemic further disrupted traditional work patterns, giving employees worldwide more flexibility in terms of where and how they work. The intersection of global megatrends alongside the COVID-19 pandemic has catalysed shifts in regional labour markets, such as an increase in remote work arrangements. Research has shown that technology-driven work arrangements, such as remote work and FWA, have a significant impact on work-life balance, productivity, and job satisfaction. Continuous monitoring of these trends, incorporating updated data, will be essential to enhance women's access to decent work and promote inclusive economic development.

Looking ahead, the integration of advanced tools and AI technologies is poised to revolutionise work-life balance, productivity, and job satisfaction for working mothers. Future trends will increasingly rely on technology to overcome geographical and institutional barriers, thereby offering flexible and supportive work environments. However, it is crucial to address the drawbacks of technology-driven work arrangements, such as social isolation, technological difficulties, and ethical considerations. Organisations must consistently evaluate and enhance their policies and practices to create truly inclusive workplaces. Harnessing technological advancements and cultivating supportive work environments will be essential in shaping future work arrangements. Visionary strategies that aim to provide equitable opportunities and enhance the well-being of working mothers will ensure their sustained participation and contribution to competitive labour markets.

Highlights

- Reimagining the future of work for women in Asia means addressing persistent structural barriers, such as limited workplace support and unpaid care responsibilities.
- Legislative reforms across the region, such as paid parental leave, are essential to foster work environments that support women's full participation.
- Future-focused businesses can play a critical role in advancing women's employment by implementing flexible work arrangements and inclusive workplace policies.

References

Arora, S., & Kumari, N. (2022). Imagining benefits and challenges for future hybrid workplace to enable reentry for women on career break. *International Journal of Networking and Virtual Organisations, 26*(3), 231–248. https://doi.org/10.1504/IJNVO.2022.122857

Armstrong, M. (2019). *Women work more than men.* Statista. https://www.statista.com/chart/13940/women-work-more-than-men/

Asian Development Bank. (2016, October). Female labor force participation in Asia: Key trends, constraints, and opportunities. https://www.adb.org/publications/female-labor-force-participation-asia

Atan, A., Ozgit, H., & Silman, F. (2021). Happiness at work and motivation for a sustainable workforce: Evidence from female hotel employees. *Sustainability (Switzerland), 13*(14). https://doi.org/10.3390/su13147778

Baird, M., Cooper, R., Parker, J., & Donnelly, N. (2023). *Women and work in Asia and the Pacific: Experiences, challenges and ways forward.* https://books.google.com

Booth, A. (2016). Women, work and the family: Is Southeast Asia different? *Economic History of Developing Regions, 31*(1), 167–197. https://doi.org/10.1080/20780389.2015.1132624

Boserup, E., Tan, S. F., & Toulmin, C. (2007). *Woman's role in economic development* (1st ed.). Routledge. https://doi.org/10.4324/9781315065892

Cameron, L. (2023). Gender equality and development: Indonesia in a global context. *Bulletin of Indonesian Economic Studies, 59*(2). https://doi.org/10.1080/00074918.2023.2229476

Chakraborty, T., Ganguly, M., & Natarajan, A. (2019). Predicting entrepreneurial satisfaction: The role of non-financial incentive factors and quality of life among women digital entrepreneurs. *Journal for Global Business Advancement, 12*(3), 328–355. https://doi.org/10.1504/JGBA.2019.101388

Desai, V. P., & D'souza, L. (2024). Women professionals in construction industry: Barriers and approaches to improve wellbeing, safety and health. *Journal of Engineering, Project, and Production Management, 14*(2), 0013. https://doi.org/10.32738/JEPPM-2024-0013

Dhanya, J. S., & Kinslin, D. (2017). Clair model for better work life balance and empowerment of women faculty in professional colleges of Kerala. *International Journal of Applied Business and Economic Research, 15*(5), 207–246.

Employment Act 1955, Malaysia. (1955). *Laws of Malaysia.* Act 265.

Fathima, F. N., Awor, P., Yen, Y. C., Gnanaselvam, N. A., & Zakham, F. (2020). Challenges and coping strategies faced by female scientists—A multicentric cross sectional study. *PLoS ONE, 15*(9), Article e0238635. https://doi.org/10.1371/journal.pone.0238635

Ganguly, A., Joseph, J. M., Dutta, S., & Dey, K. (2022). Exploring the employer–employee relationship: A management versus employee perspective of the vicissitudes in the virtual workplace. *Global Business Review, 0*(0). https://doi.org/10.1177/09721509221086353

Gideon, S. (2024, May 13). Over half of Southeast Asian working mums surveyed are adopting innovative ways of time management. *Human Resources Online*. https://www.humanresourcesonline.net/over-half-of-southeast-asian-working-mums-surveyed-are-adopting-innovative-ways-of-time-management

Government of Thailand. (2023). The Labour Protection Act (No. 8) BE 2566. *Government Gazette*.

Gupta, S., & Pathak, G. S. (2024). Ethical issues in virtual workplaces: Evidence from an emerging economy. *European Journal of Training and Development, 48*(1–2), 67–89. https://doi.org/10.1108/EJTD-03-2022-0023

Ha, C. (2024, May 9). More working mothers demand support for better work-life balance in Asia. *HRM Asia*. https://hrmasia.com/more-working-mothers-demand-support-for-better-work-life-balance-in-asia/

Ho, P. J., Sim, T. M. Y., & Loo, C. K. Y. (2024). Challenges, experiences, and potential supports for East and Southeast Asian mothers in the workforce: A systematic review. *BMC Women's Health, 24*, 422. https://doi.org/10.1186/s12905-024-03255-0

HR Asia. (2024, May 8). 6 in 10 working mothers in SE Asia struggle to find time for both work and family commitments. https://hr.asia/featured-news/6-in-10-working-mothers-in-se-asia-struggle-to-find-time-for-both-work-and-family-commitments/

International Labor Organization & Asian Development Bank. (2023). *Where women work in Asia and the Pacific: Implications for policies, equity and inclusive growth*. https://www.adb.org/sites/default/files/publication/916051/where-women-work-asia-pacific.pdf

Jain, M. (2023). Driving diversity: Nudging towards gender balance in India's workforce. *Journal of Organizational Behavior Education, 16*, 127–140. https://www.scopus.com/inward/record.uri?eid=2-s2.0-85189559200&partnerID=40&md5=716924d31a36f359d48c477bf25160fa

Japan Times. (2024, May 24). Japan passes law on flexible work for parents with young children. *The Japan Times*. https://www.japantimes.co.jp/news/2024/05/24/japan/child-care-leave-law/

Javed, H. (2024). Offering flexible work arrangements and remote work policies. In *Innovative human resource management for SMEs*. https://www.scopus.com/record/display.uri?eid=2-s2.0-85193677412&origin=resultslist&sort=plf

Kaakandikar, R., & Gawande, R. (2023). Unveiling the virtual frontier: Exploring the dynamics, challenges, and opportunities of remote work and virtual teams. *Community Practioner*, 20(7), 136–146. https://doi.org/10.17605/OSF.IO/JUPHF

Kaur, A. (2004). *Costed not valued: Women workers in industrializing Asia*. Palgrave Macmillan.

Kwek, C. L., Lee, M. S. C., Lee, S. P., Siek, H. L., Keoy, K. H., & Cherukuri, A. K. (2024). Antecedents of intention to use social media for virtual events. *Journal of Information and Knowledge Management, 23*(1). https://doi.org/10.1142/S0219649223500636

Loichinger, E., & Cheng, Y. A. (2018). Feminising the workforce in ageing East Asia? The potential of skilled female labour in four advanced economies. *Journal of Population Research, 35*(2), 187–215. http://www.jstor.org/stable/45179141

Manpower Law, Law No. 13 of 2003. (2003). *Laws of Indonesia*.

Masuda, A. D., Poelmans, S. A. Y., Allen, T. D., Spector, P. E., Lapierre, L. M., Cooper, C. L., O'Driscoll, M., Sanchez, J. I., & Kinnunen, U. (2012). Flexible work arrangements availability and their relationship with work-to-family conflict, job satisfaction, and turnover intentions: A comparison of three country clusters. *Applied Psychology: An International Review, 61*(1), 1–29. https://doi.org/10.1111/j.1464-0597.2011.00453.x

McKinsey. The future of (Hybrid) work podcast (2022). Transcript available at: https://www.mckinsey.com/featured-insights/future-of-asia/future-of-asia-podcasts/the-future-of-hybrid-work

Milieu Insight. (2024, May 10). Survey: 6 in 10 working mothers in Southeast Asia struggle with balancing work and family commitments. *Ova*. https://ova.galencentre.org/survey-6-in-10-working-mothers-in-southeast-asia-struggle-with-balancing-work-and-family-commitments/

Minister of Manpower. (2019). Regulation No. 11 of 2019 on Labor flexibility. *Laws of Indonesia*.

Ministry of Manpower. (2024, April 16). Launch of tripartite guidelines on FWA requests and tripartite workgroup report. https://www.mom.gov.sg/newsroom/press-releases/2024/0416-launch-of-tripartite-guidelines-on-fwa-requests-and-tripartite-workgroup-report

Mushfiqur, R., Mordi, C., Oruh, E. S., Nwagbara, U., Mordi, T., & Turner, I. M. (2018). The impacts of work-life balance (WLB) challenges on social sustainability: the experience of Nigerian female medical doctors. *Employee Relations, 40*(5), 868–888. https://doi.org/10.1108/ER-06-2017-0131

Naaman, S. (2022). Is telework really feasible in Israel? In *12th Griffiths school of management and IT annual conference on business, entrepreneurship and ethics, GSMAC 2022, Oradea, 21 January 2022* (pp. 169–188). Springer. https://doi.org/10.1007/978-3-031-09421-7_11

Nishimura & Asahi. (2023). Enactment of working from home law in Thailand. https://www.nishimura.com/en/knowledge/publications/enactment-of-working-from-home-law-in-thailand

Oo, B. L., & Lim, B. T. H. (2021). Changes in job situations for women workforce in construction during the COVID-19 pandemic. *Construction Economics and Building, 21*(2). https://doi.org/10.5130/AJCEB.v21i2.7526

Parker, S. (2019). *Flexible work arrangements. Women in Research.* https://www.womeninresearch.org.au/flexiblework-arrangements

Parry, A. (2024). The flexibility paradox and spatial-temporal dimensions of COVID-19 remote work adaptation among dual-earner mothers and fathers. *Gender, Work and Organization.* https://doi.org/10.1111/gwao.13130

Peslak, A., Ceccucci, W., Jones, K., & Leonard, L. (2023). Mental health and information technology developers: An exploration of demographic variances. *Issues in Information Systems, 24*(2), 250–264. https://doi.org/10.48009/2_iis_2023_122

Pfordten, D. (2024, May 12). Mums' well-being in question. *The Star*. https://www.thestar.com.my/news/nation/2024/05/12/mums-well-being-in-question.

Pillai, S. V., & Prasad, J. (2023). Investigating the key success metrics for WFH/remote work models. *Industrial and Commercial Training, 55*(1), 19–33. https://doi.org/10.1108/ICT-07-2021-0053

Reid, A. (1988). *Southeast Asia in the age of commerce 1450–1680, Volume 1: The lands below the wind*. Yale University Press.

Rodríguez-Modroño, P., & López-Igual, P. (2021). Job quality and work—Life balance of teleworkers. *International Journal of Environmental Research and Public Health, 18*(6), Article 3239. https://doi.org/10.3390/ijerph18063239

Seo, J. (J.), & Kim, E. E. K. (2024). Flexible work systems: Preparing employees for the new normal. *Journal of Business Strategy, 45*(2), 133–141. https://doi.org/10.1108/JBS-10-2022-0175

Shastri, S., Shastri, S., & Pareek, A. (2019). Motivations and challenges of women entrepreneurs: Experiences of small businesses in Jaipur city of Rajasthan. *International Journal of Sociology and Social Policy, 39*(5–6), 338–355. https://doi.org/10.1108/IJSSP-09-2018-0146

Shifrin, N. V., & Michel, J. S. (2021). Flexible work arrangements and employee health: A meta-analytic review. *Work & Stress, 36*(1), 60–85. https://doi.org/10.1080/02678373.2021.1936287

State Council Information Office. (2021, July 22). Third-child policy to unleash childbirth potential. *Gov.cn.* https://english.www.gov.cn/statecouncil/ministries/202107/22/content_WS60f8c05ac6d0df57f98dd5ef.html

Subramaniam, A. G., Overton, B. J., & Maniam, C. B. (2015). Flexible working arrangements, work life balance and women in Malaysia. *International Journal of Social Science and Humanity, 5*(1), 34.

Syed Salleh, S. N., & Mansor, N. (2022). Women and labour force participation in Malaysia. *Malaysian Journal of Social Sciences and Humanities (MJSSH), 7*(7), e001641. https://doi.org/10.47405/mjssh.v7i7.1641

Taneja, S., Pryor, M. G., & Oyler, J. (2012). Empowerment and gender equality: The retention and promotion of women in the workforce. *The Journal of Business Diversity, 12*(3), 43–53.

Teoh, W. M. Y., Teoh, Y. Y., Teoh, M. L., & Thye, W. Y. (2024). Empowering women in the workplace: Assessing female employees' readiness for flexible work arrangements. *Asian Economic and Financial Review, 14*(2), 143–153. https://doi.org/10.55493/5002.v14i2.5011

The Edge Malaysia. (2024). My say: Ten years on: What has changed for the Malaysian working woman? *The Edge Malaysia*. https://theedgemalaysia.com/node/706491

Thilagavathy, S., & Geetha, S. N. (2021). Work-life balance—A systematic review. *Vilakshan—XIMB Journal of Management*. https://doi.org/10.1108/VJIKMS-04-2021-0001

Watanabe, K. S., & Kwarteng, A. H. (2024). Unveiling the glass ceiling phenomenon and mitigating strategies through organizational justice: A conceptual paper. *Cogent Business & Management, 11*(1), Article 2331981. https://doi.org/10.1080/23311975.2024.2331981

Williamson, S., & Colley, L. (2023). Regulating for gender equality in the Australian public service: Extending Dickens' tripod of regulation. *Journal of Industrial Relations, 65*(3). https://doi.org/10.1177/00221856231 1665

World Bank. (2024, October 10). South Asia's growth outpaces expectations, more jobs for women needed to unlock full potential. https://www.worldbank.org/en/news/press-release/2024/10/10/south-asia-s-growth-outpaces-expectations-more-jobs-for-women-needed-to-unlock-full-potential.

World Economic Forum. (2018, September 10). Why Asia's future is female. https://www.weforum.org/stories/2018/09/asia-the-future-is-female/

World Economic Forum. (2023, June 20). Global gender gap report 2023: Gender gaps in the workforce. https://www.weforum.org/publications/global-gender-gap-report-2023/in-full/gender-gaps-in-the-workforce/

Yadav, M., & Jadhav-Tilekar, S. (2021). A study of work-life balance of women employees during work from home in the period of COVID-19. *International Journal of Future Generation Communication and Networking, 14*(1), 2063–2066.

Yu, R., Burke, M., & Raad, N. (2019). Exploring impact of future flexible working model evolution on urban environment, economy, and planning. *Journal of Urban Management, 8*(3). https://doi.org/10.1016/j.jum.2019.05.002

Donna Barclay obtained her legal qualifications in Australia and worked for the Attorney General's Department in Australia before relocating to Malaysia where she has been a legal educator for 20 years. She is currently a lecturer at Swinburne University of Technology Malaysia, specialising in Employment Law.

Nivakan Sritharan is a lecturer at Swinburne University of Technology Malaysia, specialises in accounting and business digitalisation. He has published extensively and is a member of the Malaysian Accounting Association.

Jacob King Soon Ting is an experienced lecturer specialising in capstone projects and MBA courses in ethics and governance. With a strong research focus in Intellectual Property Law and Commercial Law, he brings valuable academic insights to his students at Swinburne University of Technology Malaysia.

Part III
Future Workplace Technology in Asia

Chapter 7
Unlocking Career Success: How Digital Literacy Shapes Future Job Journeys

Cluster of Future Studies: Technological Trends

Poh Ling Chow, Kit Teng Phuah, Bernard Jit Heng Lim, Siti Intan Nurdiana Wong Abdullah, Mun Waei Ow, and Tai Wai Khoong

Abstract Asia's workforce is experiencing rapid changes driven by technological advancements, shifting economic priorities, and evolving skill demands. Digital transformation highlights the critical need for digital literacy and proficiency in emerging technologies. As the job market increasingly prioritizes advanced skills, particularly in management and technical roles, Malaysia envisions sustainable, productivity-driven growth supported by a digitally proficient workforce. However, research has focused mainly on education, leaving a gap in understanding how digital literacy impacts employability across industries. This study bridges that gap by examining how specific digital skills enhance job prospects. Using quantitative methods, data from 278 full-time employees were analysed through self-administered questionnaires to assess digital competencies and employment outcomes. The findings reveal that individuals with strong digital proficiency enjoy better job security, career advancement opportunities, and adaptability to industry changes. These insights inform strategies for governments, businesses, and workforce organisations

P. L. Chow (✉) · K. T. Phuah · B. J. H. Lim · M. W. Ow · T. W. Khoong
Tunku Abdul Rahman University of Management and Technology, Setapak, Malaysia
e-mail: plchow@tarc.edu.my

K. T. Phuah
e-mail: phuahkt@tarc.edu.my

B. J. H. Lim
e-mail: limjh@tarc.edu.my

M. W. Ow
e-mail: owmw@tarc.edu.my

T. W. Khoong
e-mail: khoongtw@tarc.edu.my

S. I. N. W. Abdullah
Nottingham Trent University, Nottingham, UK
e-mail: diana.abdullah@ntu.ac.uk

C. Y. Fung et al. (eds.), *Future of Work in Asia*,
https://doi.org/10.1007/978-3-031-95804-5_7

to enhance human capital development. Targeted investments in digital upskilling and integrating digital literacy into workforce training are crucial for fostering innovation, improving productivity, and ensuring competitiveness in the global digital economy. This study highlights that the workforce should be made aware of the advantages of digital technologies and government policies should emphasise the accessibility of training, raising public awareness of these advantages, and fostering an environment that encourages the adoption of digital technologies. The findings of this study have the potential to boost worker competitiveness in the global market and promote the adoption of digital technologies throughout Asia. In addition, this research provides a roadmap for aligning workforce development with curriculum design and professional growth to prepare Asia's workforce for future challenges.

Keywords Career success · Digital literacy · Employability

7.1 Introduction

The proliferation of technology and its application in the process of digitalisation has brought a heightened demand for digital literacy. The dependence on digital technology is growing quickly, and it is important to carefully analyse how people use digital technology, how they engage online, and the requisite skills they must perform for the activities related to digitalisation. Furthermore, the demand for labour qualifications in the work sector has transformed, as digital skills have become a crucial prerequisite for the long-term viability of firms, the well-being of workers, and the advancement of economies (Reddy et al., 2023). Without digital skills, new employees may not be able to sustain their economic capability and will perish in the new digital economy. The need for digital literacy in the workplace is on the rise, and in response, the education system will need to implement new strategies and programs to better prepare students for careers in the digital age (Shevyakova et al., 2021).

Consequently, new employees who lack the necessary digital skills and expertise encounter challenges in adapting to technology-based learning. Individuals are either unable to advance in their academic pursuits or withdraw from the educational system (Heber et al., 2017). Digital literacy is the remedy for reducing high attrition rates in online education and equipping graduates with the necessary skills for the modern job (Maphosa & Bhebhe, 2019). Proficiency in digital abilities is associated with higher performance and productivity in organisations is the basis for benchmarking, and people without such skills will be at a disadvantage (Anthonysamy, 2020). Furthermore, organisations are aware of the disparity between the current and required digital skills (Overlander et al., 2019). An increasing number of companies are arranging digital workshops and training sessions to guarantee that their employees possess the necessary digital skills required for the future of digitalisation (Reddy et al., 2020).

The rapid advancement of digital technologies, such as smart technology, artificial intelligence (AI), automation, robotics, cloud computing, and the Internet of Things (IoT), is profoundly reshaping the work and organisational landscape. This digitalisation's impact is evident across various job categories and industries, necessitating organisations to embrace new technology and update business models to remain competitive. Asia's digital sphere has experienced significant expansion in recent years, encompassing advancements in manufacturing automation, e-commerce platforms, and digital payment systems. Before the pandemic, the region accounted for 60% of patents in digital and computer technology, marking a notable increase from 40% two decades ago (Sayeh et al., 2023). Notably, the manufacturing sector holds a considerable global advantage in industrial robot deployment, with China being the largest consumer, representing over 30% of the market.

Figures 7.1 and 7.2 illustrate the swift technological advancements in Asian countries, fueled by the proliferation of robotic technologies, advancements in digital infrastructure R&D, and the expansion of digital payment and e-commerce platforms (Sayeh et al., 2023). Table 7.1 from the World Digital Competitiveness Ranking 2022 by the International Institute for Management Development (IMD) allows for comparisons between ASEAN countries and other key economies in Asia–Pacific. This underscores the increasing significance of digital skills certifications in Asia for enterprise sustainability, worker welfare, and economic growth.

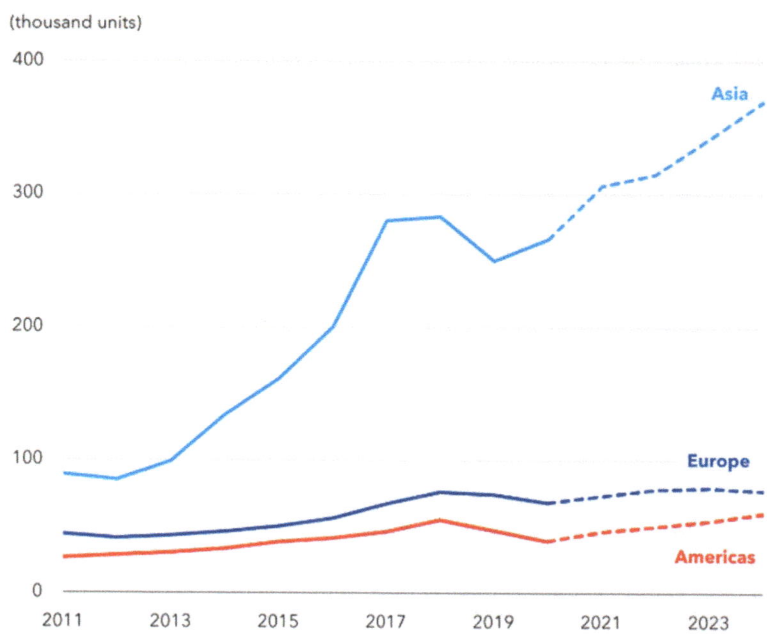

Fig. 7.1 Installation of industrial robots in Asia (reproduced from Sayeh et al., 2023)

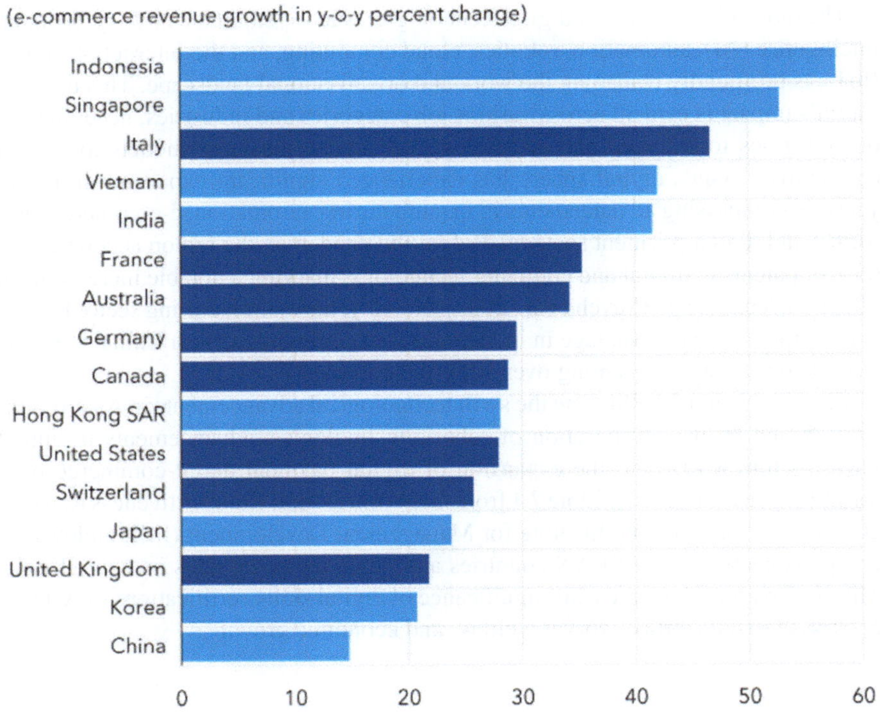

Fig. 7.2 E-commerce digitalisation trend in Asian countries (reproduced from Sayeh et al., 2023)

Table 7.1 Asian countries' talent ranking in 2022

Countries	Talent ranking (63 global countries)
Indonesia	51
Malaysia	33
Philippines	54
Singapore	12
Thailand	45
China	40
Hong Kong	14
India	52
Japan	41
Republic of Korea	38
Taiwan	19

Source Adopted from Martinus and Seth (2023)

Various educational models and frameworks have incorporated digital literacy into their structure, attempting to boost digital literacy among learners. Such a framework is summarised in the following Table 7.2.

ASEAN, the third-largest region with 400 million internet users, is a focal point for digital literacy (Agbisit & Devi, 2023). Molika (2023) notes varied digital literacy levels, with Singapore leading at 62% rating their skills as excellent, compared to 23% in Laos and 20% in Myanmar. This digital literacy gap poses a challenge in higher

Table 7.2 Various educational models and frameworks to promote digital literacy

Framework model	Description	Merit	References
The education models of the organisation for economic co-operation and development (OECD)	Involves a measure of innovations in education, change in pedagogical practices inside and outside of classrooms and the evolution of teaching practices	Include core foundations, knowledge, skills, attitudes and values, transformative competencies, and a cycle of anticipation, action, and reflection, which can be used to develop other competencies such as financial literacy, global competency, and media literacy	Ross (2019)
E3 learning model (entrepreneurship, employability, education)	Focuses on the project-based learning process for reinforcing real-life knowledge and experience, encourages learners to be critical and independent thinkers and workers, gain confidence, and become problem-solvers	To meet the goals of the new learning domains which are defined as 21st-century competencies/skills/values/ attitudes/ethics, soft skills, life skills, social-emotional skills, intra-personal and inter-personal skills, and global competencies	Basic Education, Republic of South Africa (2020)
Partnership for 21st century learning (P21)	21st-century learning model developed in collaboration with teachers, education experts, and business leaders to clearly describe, and visualise the skills and knowledge that learners need for the twenty-first century	Promotes the skills, knowledge and expertise students must master to succeed in work and life; it is a blend of content knowledge, specific skills, expertise and literacy	Remake Learning (2016)
The TPACK and SAMR models	Encourage teachers to have competencies and a good comprehension of digital concepts and theories of teaching for better learning process	Incorporate the competencies like model-technology knowledge, content knowledge and pedagogical knowledge	Alivi (2019)

Table 7.3 Percentage of young people in ASEAN countries learning digital skills at school in 2020

Country	Percentage
Vietnam	76
Philippines	71
Brunei	70
Malaysia	63
Cambodia	62
Thailand	61
Indonesia	61
Singapore	59
Laos	40
Myanmar	25

Source Adopted fromStatista (2020)

education, as self-directed learning gains popularity, impacting the significance of traditional degrees (Table 7.3). Focusing on nurturing digital skills in the youth becomes imperative.

ASEAN countries, including Malaysia, are actively enhancing digital literacy among youth, which is evident through initiatives like the Malaysian Smart School launched in 1999. This structured educational facility is dedicated to revolutionising teaching and learning, aligning with the National Philosophy of Education and the Malaysian Education Development Plan (2013–2025). The Ministry of Education utilises ICT to elevate learning value across the Malaysian education system. Presently, 75% of Malaysians possess basic ICT skills, 50% have standard skills, and 15% exhibit advanced ICT proficiency (International Telecommunication Union, 2019). To further progress digital literacy, addressing issues such as online scams, income disparity, controlled internet exposure, and facility enhancement is crucial.

In 2024, the "AI for the people" initiative underscores Malaysia's commitment to promoting both digital and AI literacy. With the digital sector contributing significantly (23.2% to GDP), the government aims to align the workforce with evolving demands. Employers witness a surge in demand for digital expertise, with a projected 7.6% growth in the next three years. Basic digital skills are deemed essential by 83.1% of employees, mirroring regional trends. High-level competencies like data analysis (49.2%) and digital marketing (47.7%) are crucial, while over 80% of organisations prioritise data analytics for expansion. The survey underscores the importance of cybersecurity (35.4%), cloud computing (35.4%), and AI/ML (20%) as significant skills (The Economist Group, 2023).

The primary aim of the study is to investigate the contribution of digital literacy to employability and future job prospects. For future workforce, acquiring sufficient digital literacy is essential for active involvement in the economy and digital society, especially considering the ongoing digital revolution in both the content and organisation of the work environment.

7.2 Literature Review

7.2.1 The Link Between Digital Literacy and Employability

An individual has the highest employability when they have broad-based education, training, initial to high-level skill sets including digital literacy, numeracy and computing capabilities, ICT skills, action planning and decision-making skills, self-awareness, self-confidence, and readiness for lifelong learning (Bejaković & Mrnjavac, 2020). Tushar and Sooraksa (2023) found that ICT skills are the essential top ten employability skills from the year 2000 until the present.

7.2.2 Self-Assessment and Identifying Skill Gaps

Evans et al. (2023) found that male students from four community colleges in the USA rated their data literacy skills higher than female students. The age group between 18 and 19 has relatively lower confidence in their data literacy scores than other age groups. High school graduates felt that they did not have proficiency in data literacy. Full-time students have higher confidence in their data literacy than part-time and unemployed students. Martzoukou et al. (2020) found that students' self-assessment of digital competencies need information literacy, digital creation, digital research, and digital identity management.

7.2.3 Digital Skills and Employability

Employability is the main result of high-quality education and training which includes abilities, competencies, knowledge, and skills that enable an individual to secure and retain a job, progress at work, adapt to changes, and secure another job in the future (Bejaković & Mrnjavac, 2020). Rahmat et al. (2022) revealed that fresh graduates with high social digital skills are highly demanded by many employers, which leads to better employability opportunities.

7.2.4 Digital Literacy (Media/Visual, Information Literacy-Refer to the Conceptual Framework)

Media/visual literacy is the proficiency in using digital technologies to access, read, interpret, analyse, evaluate and communicate information in visual presentations. Information literacy is the ability to use digital technology tools to discover, locate, analyse, evaluate, process and synthesise resources in an accurate, ethical, effective and efficient manner (Reddy et al., 2020). Consumers with high media, visual, and information literacy are more likely to have a positive attitude to use digital technology frequently, as they perceive it as both useful and easy to navigate. Nikou et al. (2022) found both digital literacy and information literacy have a positive impact towards perceived ease of use but not perceived usefulness. According to Zahoor et al. (2023), managers' digital literacy is positively related to the use of digital technologies in the context of small-medium enterprises (SMEs) operations. Besides that, Jang et al. (2020) found that information literacy positively impacts effort expectancy when using digital technologies for learning among Korean respondents. In addition, Shin and Zanuddin (2019) demonstrated that consumers with higher media literacy tend to use new media more frequently, which enhances their ability to create and share media content. Therefore, this research proposes the following hypotheses (H):

H1: Media/Visual Literacy has a positive impact towards the usage of digital technology.
H2: Information Literacy has a positive impact towards the usage of digital technology.
H3: Media/Visual Literacy has a positive impact towards perceived ease of use.
H4: Information Literacy has a positive impact towards perceived ease of use.
H6: Media/Visual Literacy has a positive impact towards perceived usefulness.
H7: Information Literacy has a positive impact towards perceived usefulness.
H10: Media/Visual Literacy has a positive impact towards attitude.
H11: Information Literacy has a positive impact towards attitude.

7.2.5 Usage of Digital Technologies

Usage of digital technologies refers to individuals' evaluation towards their usage of digital technology tools. Thompson et al. (1994) suggested that an individual's level of expertise may be correlated with the length of time they have used a personal computer. When consumers use digital technology frequently, they become more proficient in navigating it and recognise its usefulness. Liu et al. (2024) found that usage of technology positively influences individuals' cognitive benefit during leisure activities. Therefore, this research proposes the following hypotheses:

H5: Usage of digital technology has a positive impact towards perceived ease of use.
H8: Usage of digital technology has a positive impact towards perceived usefulness.

7.2.6 Perceived Ease of Use

Perceived ease of use is defined as the extent to which individuals believe that using digital technologies would be free of effort (Davis, 1989). When consumers perceive digital technologies as easy to navigate, they will recognise their usefulness and develop a positive attitude toward using them. Nikou et al. (2022) found that perceived ease of use has a positive impact towards perceived usefulness and attitude towards digital technologies in the workplace. Saksono and Untoro (2023) found that perceived ease of use has a positive effect on attitudes towards using the Shopee application. Besides that, Sayeh et al. (2023) also confirmed that perceived ease of use positively influences both perceived usefulness and attitude toward using digital technology. Therefore, this research proposes the following hypotheses:

H13: Perceived ease of use has a positive impact towards attitude.

7.2.7 *Perceived Usefulness*

Perceived usefulness is described as the extent to which individuals believe that using digital technologies would enhance job performance (Davis, 1989). When consumers recognise the usefulness of digital technologies, they are more likely to develop a positive attitude toward using them. Nikou et al. (2022) validated that perceived usefulness has a positive impact towards attitude. Shen et al. (2022) also revealed that perceived usefulness positively influences attitudes towards the application of augmented reality and virtual reality applications in tourism education. Besides that, Saksono and Untoro (2023) found that perceived usefulness has a positive effect on attitudes towards using the Shopee application in Surakarta. Therefore, this research proposes the following hypothesis:

H9: Perceived usefulness has a positive impact towards perceived ease of use
H12: Perceived usefulness has a positive impact towards attitude.

7.2.8 *Attitude to Use*

Attitude to use digital technologies refers to the extent to which individuals have a favourable or unfavourable evaluation or appraisal of using digital technologies (Ajzen, 1991). Consumers with a positive attitude toward using digital technologies are more likely to be hired by companies. Jain et al. (2023) found that attitude to use digital technologies has a positive impact towards employability. Beyond the research on attitudes toward digital technology, Jain and Jain (2013) found that a career-oriented attitude positively influences the competitive strength of Indian technology graduates' employability. Nimmi and Zakkariya (2021) found that a protean career attitude positively influences perceived employability. Therefore, this research proposes the following hypothesis:

H14: Attitude has a positive impact towards employability.

Figure 7.3 illustrates the relationship between media/visual literacy, information literacy and the usage of digital technology in shaping perception of ease of use

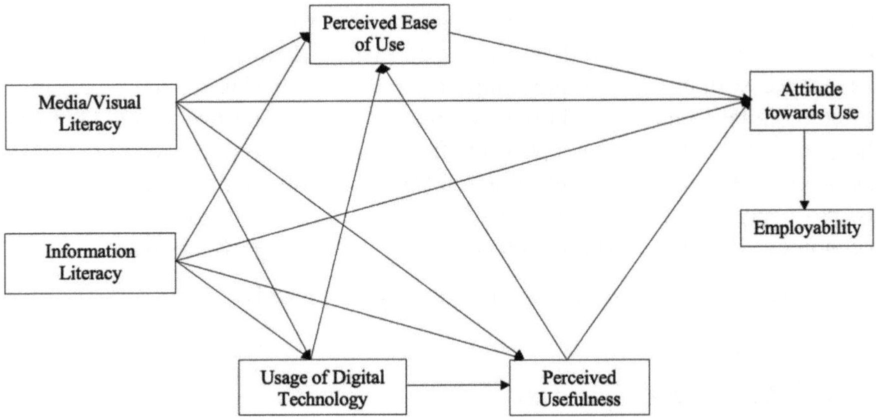

Fig. 7.3 Conceptual framework of digital literacy and employability in Malaysia. *Source* Nikou et al. (2022); Reddy et al. (2023)

and usefulness. These perceptions influence attitudes towards technology use which ultimately impacts employability.

7.3 Methodology

This study employed a quantitative survey design to explore relationships among variables. Respondents (N = 278) were Malaysian adults above 18, currently employed. Data collection utilised an online survey platform. The survey, developed through a literature review, included demographic questions and 23 items measuring dependent and independent variables. Items measuring employability, usage of digital technology, and information literacy were adopted from Zahoor et al. (2023). Information literacy items were further adapted from Ukwoma, Iwindu, and Iwundu (2016), and Zahoor et al. (2023), while media/visual literacy items were based on Simons et al. (2017) and Zahoor et al. (2023). Additionally, items measuring perceived ease of use, perceived usefulness, and attitude were adopted from Reddy et al. (2023). Covariance-based structural equation modelling (CB-SEM) was applied for analysis, using AMOS software. CB-SEM, suitable for reflective constructs, was chosen based on model characteristics. Confirmatory factor analysis (CFA) assessed the measurement model's fit, evaluating indices like chi-square, CFI, TLI, RMSEA, and SRMR. Once the measurement model was established, CB-SEM explored structural relationships. Results are detailed in the subsequent section.

Table 7.4 Demographic profile of respondents (full-time employees in Malaysia)

Variable	Percentage	Variables	Percentage	Variables	Percentage
Gender		**Industry sector**		**Education level**	
Male	34.9	Public sector	21.9	Secondary Education	1.4
Female	65.1	Private sector	73.4	Diploma	65.5
		Not-for-profit sector	4.7	Bachelor's degree	27.0
Age		Others	0.0	Master	6.1
18 to 24	3.2			Ph.D.	0.0
25 to 29	11.5	**Firm age (years)**			
30 to 34	35.3	Less than 1 year	0.4		
35 to 39	34.9	1–3	0.7		
40 to 44	11.9	4–6	6.1		
45 to 49	2.2	7–9	28.4		
50 to 54	1.1	10 and above	64.4		

7.4 Development of an Employability Prediction Model

This study aimed to develop an employability prediction model by integrating digital literacy, technology usage, perceived ease of use, perceived usefulness, and attitude. Exploratory factor analysis (EFA) with 50 respondents was conducted, followed by confirmatory factor analysis (CFA) and structural equation modelling (SEM) in AMOS with a sample of 278 Malaysians. This process validated theories, confirmed factor structure, and clarified relationships among constructs.

7.4.1 Descriptive Analysis of the Demographic Profile of Respondents

Table 7.4 shows the demographic profile of respondents. The study sample consisted of 34.9% male and 65.1% female respondents. In terms of occupational distribution, 4.7% were employed in the not-for-profit sector, 73.4% in the private sector and 21.9% in the public sector. The majority of respondents had at least a decade of work experience. Age-wise, 35.3% were between 30 and 34 years old. Educationally, 65.5% of Malaysians in the sample held at least a diploma.

7.4.2 Exploratory Factor Analysis

The promax rotation method was used for the 41-item Exploratory Factor Analysis (EFA) due to the interconnections often observed among factors in social science

Table 7.5 Exploratory factor analysis assessment criteria and results for full-time employees in Malaysia

Criteria	Requirement	Malaysia
Bartlett's test of sphericity	$p < 0.01$ (Field, 2013)	0.000
Kaiser–Meyer–Olkin (KMO)	> 0.90 is marvellous (Hutcheson & Sofroniou, 1999)	0.921
Communalities	Removed a communalities value that is less than 0.5 (Field, 2013)	Information Literacy (3 items), media and visual literacy (2 item) and perceived ease of use (4 items) were removed due to the value is less than 0.5
Total variance explained	> 50% minimum requirement (Podsakoff & Organ, 1986)	58.347
First factor's variance	< 50% threshold (Podsakoff & Organ, 1986)	15.915

research, following Costello and Osborne's recommendation (2005). The confirmation of statistical assumptions, as shown in Table 7.5, further validated this analytical approach.

7.4.3 Measurement Model Assessment and CFA

7.4.3.1 Model FI Indicators

Table 7.6 presents goodness-of-fit indices for the measurement model, adhering to benchmarks suggested by Hair et al. (2010) for structural equation modelling. Absolute fit indices, including RMSEA (0.043) and SRMR (0.035), indicate acceptable fit, supported by GFI (0.912) and AGFI (0.884). Incremental fit indices (NFI, CFI, TLI, and IFI) for Malaysia (0.900, 0.961, 0.953, and 0.962) also signify acceptable fit. Parsimony fit indices such as × 2/df (1.553), PGFI (Malaysia: 0.691), and PNFI (0.744) contribute to the positive assessment. Further examination will delve into indicator reliability, discriminant validity, convergent validity, and concept reliability to enhance the comprehensive evaluation of the model's fit.

7.4.3.2 Construct Reliability

In Malaysia, all latent variables exhibit individual Cronbach's coefficients (ranging from 0.652 to 0.890) surpassing the recommended threshold of 0.60 by Kannana and

Table 7.6 Goodness-of-Fit indices for the measurement model, including absolute, incremental and parsimonious fit measures indicating the adequacy of model fit

Name of category	Name of index	Adequate of model fit	Result (Malaysia)
Absolute fit measure	Goodness-of-fit (GFI)	> 0.90	0.912
	Adjusted goodness-of-fit index (AGFI)	> 0.80	0.884
	Root mean square error of approximation (RMSEA)	< 0.08	0.043
	Standardised root mean square residual (SRMR)	< 0.08	0.035
Incremental fit measure	Normed fit index (NFI)	> 0.80	0.900
	Comparative-fit-index (CFI)	> 0.90	0.961
	Tucker-Lewis coefficient index (TLI)	> 0.90	0.953
	The increment fit index (IFI)	> 0.90	0.962
Parsimonious fit measure	Chisq/df	1.00–5.00	1.553
	Parsimony Goodness-of-Fit Index (PGFI)	> 0.50	0.691
	Parsimony normed fit index (PNFI)	> 0.50	0.744

Tan (2005). Composite reliability (CR) values also exceed the required threshold of 0.7 set by Fornell and Larcker (1981), ensuring construct reliability standards are met. Table 7.7 confirms error-free Cronbach's alpha and CR values for all structures.

7.4.3.3 Indicator Reliability

Following the criteria outlined by Hair et al. (2013), structures with high loadings indicate a strong correlation among associated metrics. It's advisable to consider removing only loadings within the 0.4 to 0.7 range if their elimination could enhance Composite Reliability (CR) or Average Variance Extracted (AVE) values. Items with loadings below 0.40 should be excluded from the scale, as recommended by Hair et al. (2011). In this study, item loadings for Malaysia ranged from 0.417 to 0.723, surpassing the recommended threshold of 0.4 suggested by Hair et al. (2011).

7.4.3.4 Convergent Validity

Convergent validity, assessed through the Average Variance Extracted (AVE) as suggested by Hair et al. (2013), measures the correlation of a construct with other

Table 7.7 Loading, Cronbach's Alpha, CR, and AVE for the measurement model

Construct	Items	Cronbach Alpha (> 0.6)	Factor loading (> 0.4)	CR (> 0.7)	AVE (> 0.5)	Skewness	Kurtosis
Employability	EMP1	0.735	0.718	0.931	0.594	0.456	– 1.565
	EMP2		0.471			0.428	– 1.593
Attitude	ATT1	0.801	0.573	0.982	0.502	0.257	– 1.934
	ATT2		0.555			0.572	– 1.672
	ATT3		0.464			0.440	– 1.807
	ATT4		0.417			0.483	– 1.766
Perceived ease of use	PEOU6	0.759	0.608	0.977	0.513	0.369	– 1.864
	PEOU7		0.473			0.425	– 1.819
	PEOU9		0.460			0.588	– 1.655
Perceived usefulness	PU1	0.783	0.672	0.999	0.548	0.471	– 1.550
	PU2		0.509			0.369	– 1.864
	PU3		0.464			0.285	– 1.919
Information literacy	IL1	0.652	0.608	0.951	0.515	0.257	– 1.934
	IL2		0.425			0.285	– 1.919
Media/visual literacy	MVL1	0.890	0.564	0.993	0.535	0.243	– 1.941
	MVL2		0.577			– 0.522	3.609
	MVL3		0.548			0.216	– 1.953
	MVL4		0.539			0.243	– 1.941
	MVL6		0.515			0.081	– 1.994
	MVL7		0.506			0.081	– 1.994
	MVL8		0.498			0.230	– 1.947
Usage of digital technologies	UsageDT1	0.717	0.723	0.929	0.577	0.425	– 1.819
	UsageDT2		0.432			0.634	– 1.599

measures of the same construct. An AVE equal to or exceeding 0.50 indicates that the construct explains over half of the variance in its indicators, while a value below 0.50 suggests insufficient accounting for item error variation (Hair et al., 2013). In the datasets, media and visual literacy, represented by four items, were excluded due to AVE values falling below the 0.50 threshold, ensuring satisfactory construct validity in the overall model, as shown in Table 7.7. According to Hair et al (2013), the main objective of the measurement model is to assess the relationship between latent variables (constructs) and their observed indicators (measured variables). The instrument evaluates composite reliability, factor loadings, and average variance extracted (AVE) to ensure that the constructs are appropriately assessed by their indicators.

Table 7.8 Results of discriminant validity for the measurement model by Fornell-Lacker criterion

	EMP	ATT	PEOU	PU	IL	MVL	USAGE
EMP	0.594						
ATT	0.510	0.502					
PEOU	0.335	0.429	0.513				
PU	0.526	0.527	0.545	0.548			
IL	0.342	0.451	0.348	0.409	0.515		
MVL	0.476	0.517	0.394	0.495	0.735	0.535	
USAGE	0.459	0.440	0.317	0.455	0.414	0.478	0.577

Notes PEOU, perceived ease of use; PU, perceived usefulness; ATT, attitude; EMP, employability; IL, information literacy; MVL, media/visual literacy; USAGE, usage of digital technologies

7.4.3.5 Discriminant Validity

Discriminant validity analysis aims to determine whether a construct is distinct from others, capturing unique phenomena not explained by other constructs (Hair et al., 2013). Fornell and Larcker's (1981) criterion was used for this assessment. Table 7.8 displays correlations among primary constructs, indicating strong associations with their respective indicators compared to other constructs. This demonstrates robust discriminant validity (Hair et al., 2013). Additionally, the correlation between exogenous constructs is below 0.85, meeting the criteria for the overall discriminant validity of the entire model (Awang, 2014). Thus, the analysis confirms each construct's uniqueness, capturing facets not represented by others in the model.

7.4.4 Structural Model Assessment

After validating the measurement model, the structural model reveals interconnections between constructs, offering insights into their complex associations (Hair et al., 2010). It allows the assessment of empirical data alignment with theory and verification of theoretical substantiation (Hair et al., 2013). Figure 7.4, created using AMOS (version 21), illustrates the research structural model.

7.4.4.1 Hypothesis Tests

Using SEM, the correlation between independent variables and employability was investigated. The results of hypothesis testing specific to the Malaysian context are presented in Table 7.9. The study reveals a significant influence of attitude ($\beta = 0.081$, $t = 9.139$, $p < 0.01$) on employability. Notably, media/visual literacy significantly impacts perceived usefulness ($\beta = 0.205$, $t = 2.976$, $p < 0.01$), perceived ease of use ($\beta = 0.194$, $t = 2.961$, $p < 0.01$), digital literacy usage ($\beta = 0.482$, $t = 7.189$, $p < 0.01$),

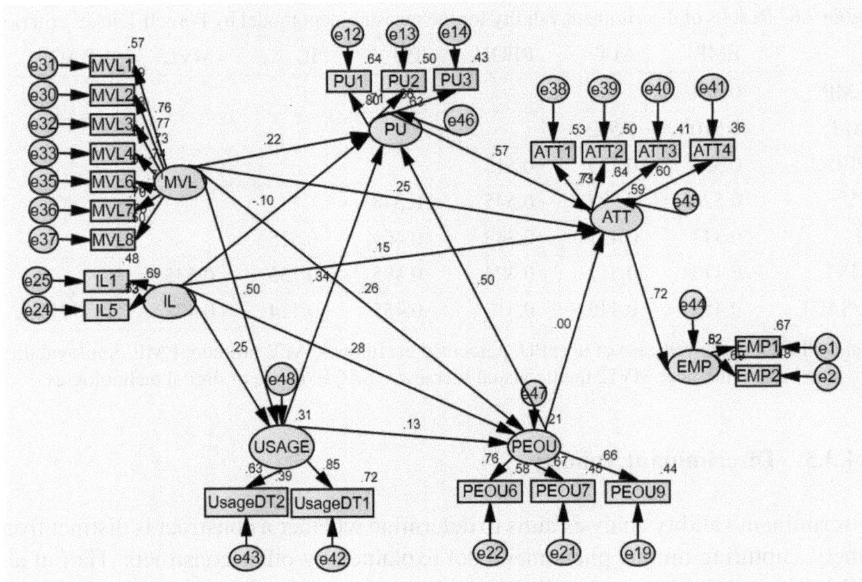

Fig. 7.4 Structural model of the measurement model for digital literacy and employability in Malaysia

and attitude (β = 0.211, t = 3.375, $p < 0.01$). This emphasises the role of media and visual literacy in fostering an understanding of functional requirements, enhancing employee performance evaluation, and positively influencing employability. Mastery of media and visual literacy is expected to positively impact perceived ease of use and usefulness in the workplace.

Our findings reveal that information literacy significantly influences perceived ease of use (β = 0.244, t = 3.078, $p < 0.01$), usage of digital literacy (β = 0.370, t = 3.098, $p < 0.01$), and attitude (β = 0.146, t = 2.122, $p < 0.01$). However, there is no significant impact of information literacy on perceived usefulness (β = − 0.104, t = − 1.338, $p > 0.05$), leading to the non-rejection of H7. As suggested by Yu et al. (2017), information literacy plays a crucial role in empowering individuals to master informational content, broaden investigative horizons, recognise information needs, and adeptly locate, evaluate, and apply necessary information.

Furthermore, the use of digital technologies (β = 0.329, t = 3.875, $p < 0.01$) significantly influences perceived usefulness. However, no significant relationship is observed with perceived ease of use (β = 0.102, t = 1.1367, $p > 0.05$), leading to the non-rejection of H5. This aligns with Oluwole's (2016) observation that individuals with higher information technology proficiency tend to perceive technology as useful in the workplace, thereby enhancing their employability.

Further analysis reveals that perceived usefulness (β = 6.402, t = 7.787, $p <$ 0.01) significantly impacts attitude. However, no significant relationship is found between perceived ease of use (β = 0.005, t = 0.048, $p > 0.05$) and attitude, leading

Table 7.9 Structural path analysis results of the measurement model for digital literacy and employability in Malaysia

	Dependent variable	Independent variable	β	S.E	C.R (t-value)	Decision
H_1	USAGE	MVL	0.482	0.067	7.189***	Supported
H_2	USAGE	IL	0.270	0.087	3.098***	Supported
H_3	PEOU	MVL	0.194	0.066	2.961***	Supported
H_4	PEOU	IL	0.244	0.079	3.078***	Supported
H_5	PEOU	USAGE	0.102	0.075	1.367	Not supported
H_6	PU	MVL	0.205	0.069	2.976***	Supported
H_7	PU	IL	− 0.104	0.078	− 1.338	Not supported
H_8	PU	USAGE	0.329	0.085	3.875**	Supported
H_9	PU	PEOU	0.614	0.101	6.050**	Supported
H_{10}	ATT	MVL	0.211	0.062	3.375**	Supported
H_{11}	ATT	IL	0.146	0.069	2.122**	Supported
H_{12}	ATT	PU	0.519	0.104	4.995***	Supported
H_{13}	ATT	PEOU	0.005	0.109	0.048	Not supported
H_{14}	EMP	ATT	0.081	0.089	9.139***	Supported

Notes PEOU, perceived ease of use; PU, perceived usefulness; ATT, attitude; EMP, employability; IL, information literacy; MVL, media/visual literacy; CL, computer literacy; CR, critical value
***Significant at 0.01, **Significant at 0.05

to the non-rejection of H13. Alamri's research (2019) highlights the importance of perceived ease of use in shaping attitudes toward adopting artificial intelligence. Table 7.9 provides valuable insights into the complex relationships between different constructs and employability in the Malaysian context.

7.5 Discussion of Implications

The findings of this study provide valuable insights into the factors influencing employability in the Malaysian context, emphasising the critical role of media/visual literacy, information literacy and digital technology usage. SEM results highlight that attitude significantly impacts employability, underscoring the importance of fostering positive attitudes among employees to enhance their job readiness. First, media/visual literacy emerges as a pivotal factor, significantly influencing perceived usefulness, perceived ease of use, digital literacy usage and attitude. These findings align with prior research (Shin & Zanuddin, 2019; Simons et al, 2017; Zahoor et al., 2023) suggesting that media and visual literacy enhance functional understanding and employee performance evaluation. For example, a Malaysian study revealed university students' moderate new media literacy, with media use weakly linked to functional and critical presumptions. The study findings underscore the urgency to enhance critical information literacy skills to combat fake news. The study highlighted that collaboration between educators and institutions is important to integrate media literacy programmes, and subsequently promote discernment in digital content consumption and creation (Shin & Zanuddin, 2019). Besides, Simons et al. (2017) developed a validated questionnaire assessing teachers' media literacy competencies, including information literacy components. The study tested 454 teachers and 219 student teachers, and findings emphasize integrating media and information literacy training to equip educators with skills to model and teach these competencies, in which strengthening media education's role in fostering critical digital engagement.

Second, information literacy also plays a significant role in shaping perceived ease of use, digital literacy usage and attitude. However, its lack of significant influence on perceived usefulness suggests that while information literacy supports the usability of digital tools, it may not directly enhance perceptions of their utility in the workplace (Jang et al., 2020; Zahoor et al., 2023). Besides, a cross-country study highlights information literacy's direct effect on digital learning intent in Korea, while digital literacy's impact is mediated by habit and performance expectancy. Findings emphasise integrating information literacy into curricula and fostering digital habits contextually, urging tailored educational strategies to bridge cultural and systemic gaps in ICT-enabled learning adoption (Jang et al., 2020).

Third, digital technology usage significantly influences perceived usefulness indicating technology proficiency enhances employability by improving perceptions of workplace relevance (Liu et al., 2024; Nikou et al., 2022). The study by Lyu and Luo (2024) demonstrates that digital literacy in China directly enhances job performance, mediated by psychological empowerment and job satisfaction. The study chain pathway digital literacy → empowerment → satisfaction → performance, which emphasisess revealed that managers' digital literacy drives digital transformation through digital technology usage, with this relationship moderated by gender, education, and age. These findings emphasise the critical role of leadership attributes in digital adoption, guiding strategies to enhance SME resilience during disruptions via context-specific managerial development (Zahoor et al., 2023). However, its non-significant relationship with perceived ease of use points to potential barriers in user experience design. Finally, the perceived usefulness strongly impacts attitude, while perceived ease of use does not. This highlights the importance of demonstrating the tangible benefits of digital tools to foster positive attitudes toward their adoption (Saksono & Untoro, 2023; Shen et al., 2022).

While existing studies highlight gaps in digital literacy research, such as geographic location focus and inconsistent operationalisation, this study underscores the urgency of equipping Malaysian students with critical digital competencies to strengthen labour markets and drive economic growth. By addressing Malaysia's unique challenges such as moderate media literacy and infrastructure barriers, this research provides a roadmap for Asia to align workforce development with curriculum design and professional growth. Hence, these findings provide several practical implications for policymakers and organisations aiming to enhance employability in Malaysia:

1. Promoting media/visual literacy—given its significant impact on multiple constructs related to employability, integrating media and visual literacy training into educational curricula and workplace development programs is essential for fostering a skilled workforce.
2. Enhancing information literacy—policymakers should prioritise initiatives that improve individuals' ability to locate and apply information effectively while addressing gaps in translating information literacy into perceived usefulness.
3. Focusing on digital technology adoption—organisations should invest in user-friendly digital tools and training programs to maximize the perceived usefulness of technology and address usability challenges.
4. Encouraging positive attitudes—employers can design interventions that highlight the practical benefits of workplace technologies to improve employee attitudes toward their adoption.
5. Developing comprehensive policies—governments should implement forward-thinking policies that integrate digital education and workforce training to ensure long-term economic resilience and competitiveness.

By addressing these areas strategically, Malaysia can cultivate a future-ready workforce with the skills necessary to thrive in an evolving global economy. Emphasising contextualised strategies such as public–private partnerships, it supports harmonising digital skill-building with regional socio-economic goals. This subsequently ensures a future-ready workforce capable of navigating technological disruptions and fostering sustainable development across diverse Asian economies. While these recommendations address current issues, their long-term success depends on anticipating emerging challenges in the digital landscape. Hence, the explores these evolving priorities, including data privacy, cybersecurity, and AI-driven skill adaptation, which will define the next phase of workforce readiness next phase.

7.6 Future Directions

Although this study can identify pathways to enhance employability, it also reveals persistent challenges, particularly in bridging the information literacy and perceived workplace utility gap. To sustain progress, it needs to look beyond foundational skills and prioritise adaptive competencies for a digitiszed future. Thus, issues such as cybersecurity literacy, AI integration, and lifelong learning frameworks, are essential to ensure the workforce remains resilient amidst increasing technological change.

The evolution of digital literacy involves safeguarding data privacy, enhancing critical thinking, and strengthening cybersecurity measures (Blackmon & Major, 2023; Sulintang et al., 2024; Weninger, 2022; Yuhertiana et al., 2024) The integration of AI and blockchain contributes significantly to providing inclusion and access to digital skills for all. First, understanding data privacy entails awareness of online data gathering, privacy configurations, scam identification, and encrypted communication (Bajunaied et al., 2023; Blackmon & Major, 2023; Prastyanti & Sharma,

2024). This emphasise on data privacy legislation such as the General Data Protection Regulation (GDPR) framework, educates individuals in regards to their digital rights and responsibilities.

Second, critical thinking in the future digital age encompasses assessing online information, distinguishing reliable sources, and making informed judgments (Ru-Zhue et al., 2025; Yuhertiana et al., 2024; Yustina et al. 2022). The focus is on questioning, analysing, and verifying digital content, fostering an analytical mindset in the vast online environment (Haryanto et al., 2022; Ru-Zhue et al., 2025). Third, digital literacy in cybersecurity includes knowledge and skills for individual protection online, covering cyber threats, scam identification, robust password use, and security protocols awareness (Alsharida et al. 2023; Diepeveen & Pinet, 2023; Senarak, 2024). Proficiency in these aspects is crucial due to the increasing frequency of cyberattacks.

Besides, proficiency in digital literacy is essential in the future job market, requiring competence in digital tools, data analysis, coding, and online collaboration (Hajizadeh et al., 2024; Zhang et al., 2024). Individuals seek adaptable candidates with strong problem-solving abilities to navigate the digital domain effectively (Almassri & Zaharudin, 2023; Nauli et al., 2024). Continuous learning through online courses, workshops, and digital tools ensures individuals maintain high digital literacy throughout their careers. Hence, a lifelong strategy for digital literacy is needed (Chatwattana et al., 2024; Mutia et al., 2024; Serafimova et al., 2024), consisting of divergence, convergence, and transformation stages. Divergence includes acquiring digital skills and exploring various instruments, technologies, and platforms. Then, convergence consolidates diverse skills for a comprehensive understanding, enabling individuals to address digital challenges holistically. Finally, transformation involves adapting and developing skills to meet evolving digital requirements, ensuring relevance in a dynamic technological landscape, such as embracing emerging technologies and shifting methodologies in response to industry trends (Mutia et al., 2024; Serafimova et al., 2024).

7.7 Conclusion

This study explores the impact of literacy dimensions and technology use on employability in Malaysia. Key findings highlight attitude's significant influence on employability ($\beta = 0.081$, $t = 9.139$, $p < 0.01$). Media/visual literacy plays a crucial role in shaping perceived usefulness, ease of use, digital literacy usage, and overall attitude, contributing to functional understanding and employability.

Information literacy significantly influences perceived ease of use, digital literacy usage, and attitude, empowering individuals to master, expand, and apply knowledge. Digital technology utilisation impacts perceived usefulness, enhancing employability. Then, perceived usefulness significantly influences attitude, emphasising the importance of user-friendly technology for positive workplace adoption. The study contributes theoretically, emphasising the complex interplay between attitude, literacy dimensions, technology use, and employability.

Practically, organisations and educational institutions in Malaysia should prioritise media/visual literacy development, leading to a more skilled workforce and improved employability. Recognising information literacy's empowering role, training programs can enhance employees' ability to effectively apply information. Promoting digital technology proficiency positively affects technology's usefulness in the workplace, enhancing employability. Besides, the understanding of the significance of perceived usefulness in shaping attitudes underscores the need for user-friendly technology adoption in the workplace.

However, the study limitations include the context-specific nature of the study in Malaysia, cautioning against generalising findings to different cultural or organisational contexts. Future research could explore these relationships in diverse settings, employing a mix of qualitative and quantitative methodologies for a comprehensive understanding.

Highlights

- Media/visual literacy enhances perceived usefulness, ease of use, digital literacy, and employability, shaping the future workforce.
- Information literacy empowers knowledge mastery, application, and attitude, shaping the future of work.
- Effective digital technology utilisation increases perceived usefulness, promoting workforce integration in an evolving job market.
- Perceived usefulness fosters positive attitudes, promoting workplace technology adoption in future.
- Practical focus: Develop literacy dimensions for a skilled, employable workforce across Asia.

References

Agbisit, J. B., & Devi, I. T. (2023, December 19). Ending misinformation through digital literacy. *The ASEAN Magazine*. https://theaseanmagazine.asean.org/article/ending-misinformation-through-digital-literacy/

Ajzen, I. (1991). The theory of planned behaviour. *Organisational Behavior and Human Decision Processes, 50*, 179–211.

Alamri, M. M., et al. (2019). Towards adaptive e-learning among university students: By applying technology acceptance model (TAM). *International Journal of Engineering and Advanced Technology, 8*(6).

Alivi, J. (2019). A review of TPACK and SMAR models: How should language teachers adopt technology? *Journal of English for Academic and Specific Purposes, 2*(1), 1–12. https://doi.org/10.18860/jeasp.v2i2.7944

Almassri, M. A. H., & Zaharudin, R. (2023). Effectiveness of flipped classroom pedagogy in programming education: A meta-analysis. *International Journal of Instruction, 16*(2), 267–290. https://doi.org/10.29333/iji.2023.16216a

Alsharida, R. A. Al-rimy, Bander, A. S. Al-Emran, M., & Zainal, A. (2023). A systematic review of multi perspectives on human cybersecurity behaviour. *Technology in Society, 73*, art no. 102258. https://doi.org/10.1016/j.techsoc.2023.102258

Anthonysamy, L. (2020). Digital literacy deficiencies in digital learning among undergraduates. In *Understanding digital industry* (pp. 133–136). Routledge.

Awang, Z. (2014). *Structural equation modeling using AMOS*. University Teknologi MARA Publication Center.

Basic Education, Republic of South Africa. (2020). E3 learning model pedagogical paradigms. https://www.ecubed-dbe.org/wp-content/uploads/2020/07/E3-Learning-Model-pedagogical-content_display.01.20.pdf

Bajunaied, K., Hussin, N., & Kamarudin, S. (2023). Behavioral intention to adopt FinTech services: An extension of unified theory of acceptance and use of technology. *Journal of Open Innovation: Technology, Market, and Complexity, 9*(1), art no. 100010. https://doi.org/10.1016/j.joitmc.2023.100010

Bejaković, P., & Mrnjavac, Z. (2020). The importance of digital literacy on the labour market. *Employee Relations: THe International Journal, 42*(4), 921–932. https://doi.org/10.1108/ER-07-2019-0274

Blackmon, S. J., & Major, C. H. (2023). Inclusion or infringement? A systematic research review of students' perspectives on student privacy in technology-enhanced, hybrid and online courses. *British Journal of Educational Technology, 54*(6), 1542–1565. https://doi.org/10.1111/bjet.13362

Chatwattana, P., Wongwiwat, W., & Suwankan, P. (2024). The learning outcomes of total learning experience (TLE) to support digital empathy and literacy skills. *Global Journal of Engineering Education, 26*(1), 26–34.

Costello, A. B., & Osborne, J. W. (2005). Best practices in exploratory factor analysis: Four recommendations for getting the most from your analysis. *Practical Assessment, Research and Evaluation, 10*(7), 1–9.

Davis, F. D. (1989). Perceived usefulness, perceived ease of use, and user acceptance of information technology. *MIS Quarterly: Management Information Systems, 13*(3), https://doi.org/10.2307/249008

Diepeveen, S., & Pinet, M. (2023). User perspectives on digital literacy as a response to misinformation. *Development Policy Review*. https://doi.org/10.1111/dpr.12671

Evans, S. A., Hong, L., Kim, J., Rice-Oyler, E., & Ali, I. (2023). Community college students' self-assessment of data literacy: Exploring differences amongst demographic, academic, and career characteristics. *Information and Learning Sciences*. https://doi.org/10.1108/ILS-06-2023-006

Field, A. (2013). *Discovering statistics using IBM SPSS statistics* (4th ed.). Sage Publications Ltd.

Fornell, C., & Larcker, D. F. (1981). Evaluating structural equation models with unobservable variables and measurement error. *Journal of Marketing Research, 18*(1), 39–50. https://doi.org/10.2307/3151312

Hair, J. F., Black, W. C., Babin, B. J., & Anderson, R. E. (2010). *Multivariate data analysis* (7th ed.). Pearson.

Hair, J. F., Hult, G. T. M., Ringle, C. M., & Sarstedt, M. (2013). *A primer on partial least squares structural equation modeling (PLS-SEM)*. Sage Publications.

Hair, J. F., Ringle, C. M., & Starstedt, M. (2011). PLS-SEM: Indeed a silver bullet. *Journal of Marketing Theory and Practice, 19*(2), 139–151. https://doi.org/10.2753/MTP1069-6679190202

Hajizadeh, S., Ebadi, S., Salman, A. R., & Badiozaman, I. F. B. A. (2024). Young twins' journey via digital storytelling towards multi-literacy development. *Journal of Visual Literacy, 43*(3), 214–232. https://doi.org/10.1080/1051144X.2024.2394333

Haryanto, H., Ghufron, A., Suyantiningsih, S., & Kumala, F. N. (2022). The correlation between digital literacy and parents' roles towards elementary school students' critical thinking. *Cypriot Journal of Educational Sciences, 17*(3). https://doi.org/10.18844/cjes.v17i3.6890

Heber, E., Ebert, D., Lehr, D., Cuijpers, P., Berking, M., Nobis, S., & Riper, H. (2017). The benefit of web- and computer-based interventions for stress: A systematic review and meta-analysis. *Journal of Medical Internet Research, 19*(2), 1–32. https://doi.org/10.2196/jmir.5774

Hutcheson, G. D., & Sofroniou, N. (1999). *The multivariate social scientist.* Sage.

International Telecommunication Union. (2019). Measuring digital development: Facts and figures 2019. https://www.itu.int/en/ITU-D/Statistics/Documents/facts/FactsFigures2019.pdf

Jain, A., Khan, S., Bhanot, S., & Bendigeri, R. S. (2023). A study on impact of technology on knowledge, skills and attitude (KSA) of students and its implications on their placements with specific reference to management institutes in Mumbai. *Journal of Research Administration, 5*(2), 7014–7024.

Jain, R., & Jain, S. (2013). Conceptualization, measure development and empirical assessment of career oriented attitudes and employability of technology graduates. *Sage Publication, 17*(2), 143–157. https://doi.org/10.1177/0972262912483528

Jang, M., Aavakare, M., Kim, S., & Nikou, S. (2020, June 14–17). The effects of digital literacy and information literacy on the intention to use digital technologies for learning—A comparative study in Korea and Finland. ITS Online Event. International Telecommunications Society (ITS).

Kannana, V. R., & Tan, K. C. (2005). Just in time, total quality management, and supply chain management: Understanding their linkages and impact on business performance. *Omega, 33*(2), 153–162. https://doi.org/10.1016/j.omega.2004.03.012

Liu, C., Horng, J., Chou, S., Yu, T., Huang, Y., & Vu, H. Y. (2024). The roles of cognitive benefits and technology usage in predicting students' sustainable behavioural intentions: The moderating roles of data literacy and marketing strategy. *Journal of Hospitality, Leisure, Sport & Tourism Education, 34,* Article 100480. https://doi.org/10.1016/j.jhlste.2024.100480

Lyu, Y., & Luo, J.-h. (2024). The relationship between digital literacy and job performance of university teachers: The chain mediating effect of psychological empowerment and job satisfaction. *Current Psychology, 43,* 34496–34510. https://doi.org/10.1007/s12144-024-06971-w

Maphosa, C., & Bhebhe, S. (2019). Digital literacy: A must for open distance and e-learning (ODEL) students. *European Journal of Education Studies, 5*(10), 186–199. https://www.oapub.org/edu/index.php/ejes/article/view/2274/4913

Martinus, M., & Seth, F. N. (2023, June 21). Current key drivers of ASEAN integration: Digital skills and mobilities. Researchers At Iseas—Yusof Ishak Institute Analyse Current Events. https://www.iseas.edu.sg/wp-content/uploads/2023/05/ISEAS_Perspective_2023_48.pdf

Martzoukou, K., Fulton, C., Kostagiolas, P., & Lavranos, C. (2020). A study of higher education students' self-perceived digital competences for learning and everyday life online participation. *Journal of Documentation, 76*(6), 1413–1458.

Molika, H. (2023, February 16). ASEAN needs to prepare now for the future of work. *The Diplomat.* https://thediplomat.com/2023/02/asean-needs-to-prepare-now-for-the-future-of-work/

Mutia, F., Masrek, M. N., Baharuddin, M. F., Shuhidan, S. M., Soesantari, T., Yuwinanto, H. P., & Atmi, R. T. (2024). An exploratory comparative analysis of librarians' views on AI support for learning experiences, lifelong learning, and digital literacy in Malaysia and Indonesia. *Publications, 12*(3), art no. 21. https://doi.org/10.3390/publications12030021

Nauli, N., Harisman, Y., & Armiati, Y. (2024). Junior high school students' habits of mind in solving mathematical problems. *Mathematics Education Journal, 18*(1), 39–58. https://doi.org/10.22342/jpm.v18i1.pp39-58

Nimmi, P. M., & Zakkariya, K. A. (2021). Determinants of perceived employability in current career context: Evidence from India. *Journal of Asia Business Studies, 15*(5), 804–820. https://doi.org/10.1108/JABS-05-2020-0187

Nikou, S., De Reuver, M., & Kanafi, M. M. (2022). Workplace literacy skills—How information and digital literacy affect adoption of digital technology. *Journal of Documentation, 78*(7), 371–391.

Oluwole, D. (2016). Technology acceptance model as a predictor of using information system to acquire information literacy skills. *Library Philosophy & Practice (e-Journal), 1450*, 1–27.

Overlander, M., Beinicke, A., & Bipp, Y. (2019). Digital competencies: A review of the literature and applications in the workplace. *Computers & Education, 146*, 1–23. https://www.sciencedirect.com/science/article/abs/pii/S0360131519303057

Podsakoff, P. M., & Organ, D. W. (1986). Self-reports in organisational research: Problems and prospects. *Journal of Management, 12*(4), 531–544. https://doi.org/10.1177/014920638601200408

Prastyanti, R. A., & Sharma, R. (2024). Establishing consumer trust through data protection law as a competitive advantage in Indonesia and India. *Journal of Human Rights, Culture and Legal System, 4*(2), 354–390. https://doi.org/10.53955/jhcls.v4i2.200

Rahmat, A. M., Mohd, I. H., Omar, M. K., Kamalludeen, R., Wan Zahari, W. M. Z., Azmy, N., & Mohd Adnan, A. H. (2022). Integrating socio-digital skills in the Industry 4.0 era for graduates' employability: An employers' perspective. *Journal of Positive School Psychology, 6*(3), 8493–8507.

Reddy, P., Chaudhary, K., & Hussein, S. (2023). A digital literacy model to narrow the digital literacy skills gap. *Heliyon, 9*, Article e14878. https://doi.org/10.1016/j.heliyon.2023.e14878

Reddy, R., Sharma, B., & Chaudhary, K. (2020). Digital literacy: A review of literature. *International Journal of Technoethics (IJT), 11*(2), 65–94. https://www.igi-global.com/article/digital-literacy/258971

Remake Learning. (2016). Demystifying learning frameworks: The P21 framework. https://remakelearning.org/blog/2016/04/29/demystifying-learning-frameworks-the-p21framework/#:~:text=The%20P21%20Framework%20is%20a,education%20such%20as%20the%20NEA

Ross, D. (2019). OECD Learning Compass 2030 provides teachers, students, and employers with milestones for workforce success. https://www.gettingsmart.com/2019/06/oecd-learning-compass-2030-provides-teachers-students-and-employers-with-milestones-for-workforce-success/

Ru-Zhue, J., Imjai, N., Yordudom, T., Kongrode, J., Sumiati, S., & Aujirapongpan, S. (2025). How do the role of digital connectivity capabilities and heuristic thinking skills shape the development of digital marketing competency in Gen Z marketing students? *International Journal of Management Education, 23*(2), art no. 101157.

Saksono, A. S., & Untoro, W. (2023). Consumer perceived ease of use and consumer perceived usefulness in using the Shopee application in Surakarta with discount as a moderation variable. *European Journal of Business and Management Research, 8*(4), 13–19. https://doi.org/10.24018/ejbmr.2023.8.4.2022

Sayeh, A. M., Dabla-Norris, E., & Kinda, T. (2023). Asia's productivity needs a boost that digitalization can provide. *IMF Blog*. https://www.imf.org/en/Blogs/Articles/2023/01/09/asias-productivity-needs-a-boost-that-digitalization-can-provide

Senarak, C. (2024). Port cyberattacks from 2011 to 2023: A literature review and discussion of selected cases. *Maritime Economics and Logistics, 26*(1), 105–130. https://doi.org/10.1057/s41278-023-00276-8

Serafimova, D., Jaroenwanit, P., & Phuensane, P. (2024). Digital age pedagogy: How European and Asian business school students perceive competence-oriented education. *Changing Societies and Personalities, 8*(4), 1006–1033. https://doi.org/10.15826/csp.2024.8.4.310

Shen, S., Xu, K., Sotiriadis, M., & Wang, Y. (2022). Exploring the factors influencing the adoption and usage of augmented reality and virtual reality applications in tourism education within the context of COVID-19 pandemic. *Journal of Hospitality, Leisure, Sport & Tourism Education, 30*, Article 100373. https://doi.org/10.1016/j.jhlste.2022.100373

Shevyakova, A., Munsh, E., Arystan, M., & Petrenko, Y. (2021). Competence development for Industry 4.0: Qualification requirements and solutions. *Insights into Regional Development, 3*(1), 124–135.

Shin, C. Y., & Zanuddin, H. (2019). New media literacy and media use among university students in Malaysia. *International Journal of Engineering and Advanced Technology (IJEAT), 8*(5C). https://doi.org/10.35940/ijeat.E1066.0585C19

Simons, M., Meeus, W., & T'sas, J. (2017). Measuring media literacy for media education: Development of a questionnaire for teachers' competencies. *Journal of Media Literacy Education, 9*(1), 99–115.

Statista. (2020). Share of young people learning digital skills at school in ASEAN countries. https://www.statista.com/statistics/1248016/asean-digital-literacy-education-at-school-by-country/

Sulintang, A., Saragih, T. S., Pramasanti, A. N., Maha-Ganti, F. S., Fitriani, K., Augustin, E., Putra, M. A., & Kara, S. B. (2024). Bibliometric analysis of Indonesia's labor dynamics: Future works, digital transformations, and contemporary employment landscape shifts. *International Journal of Data and Network Science, 8*(2), 679–688.

The Economist Group. (n.d.). Bridging the skills gap: Fuelling careers and the economy in Malaysia. https://impact.economist.com/perspectives/sites/default/files/skilling_in_asia-_malaysia_pdf.pdf

Thompson, R. L., Higgins, C. A., & Howell, J. M. (1994). Influence of experience on personal computer utilization: Testing a conceptual model. *Journal of Management Information Systems, 11*, 167–187.

Tushar, H., & Sooraksa, N. (2023). Global employability skills in the 21st-century workplace: A semi-systematic literature review. *Heliyon, 9*, Article e21023.

Ukwoma, S., Iwundu, N. E., & Iwundu, I. E. (2016). Digital literacy skills possessed by students of UNN, implications for effective learning and performance: A study of the MTN Universities Connect Library. *New Library World, 117*(11/12), 702–720. https://doi.org/10.1108/NLW-08-2016-0061

Weninger, C. (2022). Skill versus social practice? Some challenges in teaching digital literacy in the university classroom. *TESOL Quarterly, 56*(3), 1016–1028.

Yu, T. K., Lin, M. L., & Liao, Y. K. (2017). Understanding factors influencing information communication technology adoption behaviour: The moderators of information literacy and digital skills. *Computers in Human Behavior, 71*, 196–208.

Yuhertiana, I., Izaak, W. C., Rahmawati, A., & Sucahyati, D. (2024). Creative performance of lecturers in post-pandemic COVID 19: Evidence from Indonesia. *Cogent Education, 11*(1), art no. 2369974.

Yustina, Mahadi, I., Daryanes, F., Alimin, E., & Nengsih, B. (2022). The effect of problem-based learning through blended learning on digital literacy of eleventh-grade students on excretory system material. *Jurnal Pendidikan IPA Indonesia, 11*(4). https://doi.org/10.15294/jpii.v11i4.38082

Zahoor, N., Zopiatis, A., Adomako, S., & Lamprinakos, G. (2023). The micro-foundations of digitally transforming SMEs: How digital literacy and technology interact with managerial attributes. *Journal of Business Research, 159*, Article 113755. https://doi.org/10.1016/j.jbusres.2023.113755

Zhang, M., Zhang, X., Chen, Z., Wang, Z., Liu, C., & Park, K. (2024). Charting the path of technology-integrated competence in industrial design during the era of Industry 4.0. *Sustainability (Switzerland), 16*(2), art no. 751. https://doi.org/10.3390/su16020751

Poh Ling Chow is the Assistant Professor and Deputy Dean at the Faculty of Accountancy, Finance, and Business, Tunku Abdul Rahman University of Management and Technology. A Fellow of the Chartered Institute of Management Accountants (CIMA), she serves as a CIMA Academic Champion (2023–2025), Co-opted Member of the Education Training Committee (2022–2026), and Exco Member of the ASEAN Education Workgroup (2023–2026). With over 13 years of corporate experience, she is also a HRDF Certified Trainer. Her research focuses on quality education, soft skills enhancement, gender equality, and social sustainability, with contributions to academic publications and collaborative research projects.

Kit Teng Phuah is an Associate Professor at Tunku Rahman University of Management and Technology, excels in research and marketing with a focus on food-related marketing and consumer behavior. Her expertise spans agribusiness, simulation, eco-tourism, and data analytics. With numerous journal publications and significant roles in research grants, she has received the Emerald Literati Award for Most Outstanding Paper in 2018. Dr. Phuah has led more than 20 research and consultancy projects, bridging academic knowledge with real-world applications, reinforcing her status as a thought leader in her field.

Bernard Jit Heng Lim is a dedicated lecturer at Tunku Abdul Rahman University of Management and Technology, specialising in finance. Currently pursuing a Ph.D. at Putra Business School in UPM, his research interests encompass personal finance, business finance, and behavioural studies. Bernard is passionate about teaching finance-related subjects, aiming to equip students with essential knowledge and skills. He believes in the transformative power of education and is committed to inspiring young minds to reach their full potential. Through his academic endeavours, Bernard seeks to bridge theoretical concepts with practical applications, fostering a deeper understanding of finance among his students.

Siti Intan Nurdiana Wong Abdullah is a senior lecturer at the Nottingham Business School, Nottingham Trent University, UK. Her areas of research include sustainable tourism, green consumption behaviour, consumer attitudes, and sustainability in higher education. As part of her current role as a dedicated educator and researcher, Dr. Diana is a member of the Marketing and Consumer Studies (MACS) Research Group, the Chartered Institute of Marketing and a Certified Management and Business Educator (CMBE), along her active involvement in the British Academy's Early Career Researcher Network. She writes and publishes articles, participates in international conferences, serves as a reviewer, editorial member and has won numerous research grants.

Mun Waei Ow is an Associate Professor in Marketing at Tunku Abdul Rahman University of Management and Technology, is dedicated to shaping future business leaders. She specialises in business research methods and consumer behaviour, focusing on loyalty, satisfaction, attitudes, and perceptions. Her expertise spans data analytics and innovative work-integrated learning pedagogy. Dr. Ow bridges academia and industry by mentoring students in real-world marketing projects, delivering essential skills and actionable insights. A prolific researcher and contributor to journals, she excels in leading research initiatives. Her unwavering commitment to lifelong learning drives her pursuit of academic excellence and professional growth.

Tai Wai Khoong is a practitioner turned academician. He has more than 12 years of experience in finance and business development. Currently, he serves as Associate Professor at Tunku Abdul Rahman University of Management and Technology. He holds Chartered Islamic Finance Professional and Registered Financial Planner professional designations. Besides, he is a certified trainer and examiner for the Malaysian Financial Planning Council (MFPC) and serves as a committee board member of Certification & CPD Board for MFPC. He frequently being invited contribute articles to magazines and speaker. His areas of research include financial inclusion-literacy and business analytics.

Chapter 8
The Future of Work with Advanced Computer-Assisted Audit Tools and Techniques

Clusters of Future Studies: Technological Trends; Corporate Foresight

Suhaily Shahimi, Dayana Jalaludin, and Ravinder Kaur Harminder Singh

Abstract The audit industry in Asia is undergoing a transformative phase marked by the increasing utilisation of advanced computer-assisted audit tools and techniques (CAATTs). Despite this trend, there remains a dearth of discussion on the prospects of CAATTs for enhancing auditing practises and outcomes in the digital era. This conceptual article aims to address this gap by firstly spotlighting recent technological advancements that can be integrated into CAATTs software and methodologies. Secondly, it proposes the CAATTs adoption framework tailored to the fourth industrial revolution (IR4.0), encompassing technological, organisational, and environmental considerations. The next segments delve into the psychological dimension of CAATTs adoption, recognising its pivotal role in the uptake of any technology. This is then followed by implications and future directions of its adoption. The chapter concludes by outlining the foundational elements of its adoption for the Asian landscape, emphasising the importance of addressing existing barriers, fostering organisational support and motivation, encouraging auditor engagement, and customising approaches to suit regional contexts. By acknowledging both technical barriers and psychological factors, the chapter explores strategies for the swift and effective implementation of advanced CAATTs. Ultimately, this contribution offers insights beneficial for audit firms in formulating future work strategies to enhance auditing quality.

S. Shahimi · D. Jalaludin (✉) · R. K. H. Singh
Accounting Department, Faculty of Business and Economics, Universiti Malaya, Kuala Lumpur, Malaysia
e-mail: dayana@um.edu.my

S. Shahimi
e-mail: suhaily@um.edu.my

C. Y. Fung et al. (eds.), *Future of Work in Asia*,
https://doi.org/10.1007/978-3-031-95804-5_8

Keywords Audit industry · Psychological factors · TOE framework · Asia · IR. 4.0

8.1 Introduction

In response to the rapid technological advancements and the accelerating pace of digitalisation, the implementation of dynamic audits utilising computer-assisted audit tools and techniques (CAATTs) has become imperative for ensuring the efficacy of financial statement audits, assurance procedures, tax services, and other related audit services (Liew et al., 2022). Advanced CAATTs encompass a spectrum of technological tools and methodologies employed to facilitate audit processes, incorporating features such as task automation, database utilisation, visualisation techniques, robotic process automation, artificial intelligence, machine learning, and the utilisation of electronic audit documentation, often integrating data analytics, cloud-based applications, and blockchain auditing solutions. The integration of the state-of-the-art CAATTs, such as Tableau and Power BI, equips auditors with the capability to analyse extensive datasets comprehensively, enabling in-depth scrutiny and fostering the production of audit outcomes that conform to elevated standards (AICPA, 2023).

Audit tools in the era of the fourth industrial revolution (IR 4.0) are shaped by technological advancements, encapsulating a myriad of innovative features. The symbiosis of these audit technologies with the acumen and experience of auditors liberates auditing practices from labour-intensive manual tasks and disparate audit tools, transitioning towards a more cohesive and efficient audit methodology that prioritises substantive auditor–client dialogues (AICPA, 2023). Table 8.1 presents key features of technological innovations that enhance CAATTs applications when integrated into relevant audit software and methodologies.

However, the widespread adoption of these diverse technologies appears to lag both its potential and, arguably, its imperative role in contemporary audit practices. A survey conducted among members of the Association of Chartered Certified Accountants (ACCA) in March 2023, as documented by ACCA (2023), unveils a spectrum of acceptance and experimentation concerning these technologies within the business landscape. Certain technological avenues remain largely unexplored, as illustrated in Fig. 8.1, thus accentuating the disjunction between the available technological repertoire and its tangible implementation in auditing practices.

The audit industry has evolved significantly across much of Asia, with auditors now expected to meet the demands of a rapidly changing business environment (Handoyo, 2024). The recent health and economic crises, the digital revolution, and the transformation of business models, alongside new policies emphasising environmental, social, and governance (ESG) agendas, have heightened compliance demands and introduced unforeseen risks. Consequently, auditors must upskill and broaden their work scope to enhance efficiency. For instance, leveraging cloud-based technology and integrating artificial intelligence can improve risk identification and

Table 8.1 The main feature of technological advancements that may be integrated into Computer Assisted Audit Tools and Techniques (CAATTs)

Technology	Main feature
Internet of Things (IoT)	Network of physical world embedded in technology to interact, connect, and exchange data
Big Data Analytics (BDA)	Information processing of big and large set of complex data that enhances decision making, data analytics and process automation
Autonomous robots	Self-governing machine capable of performing human tasks using sensors and AI to interact with its environment
Digital Twin	Replica of a physical system, used for simulation and analysis
Cloud computing	Data storing and management allowing on-demand and real-time communication using enabling technology
Artificial Intelligence (AI)	Encompasses systems that perform complex tasks associated with human intelligence, but with superior capacity
Blockchain	Digital record with entries that are concatenated chronologically and whose integrity is protected by cryptography
Augmented reality	Technology between virtual reality and telepresence, allowing interaction of an actual reality in a synthetic environment
Machine learning (ML)	A branch of AI focusing on creating systems that can learn from data, see patterns, and make decisions with little to no human input
Cybersecurity/Cyber-Physical Systems (CPS)	Integration of visual and real world; monitoring and collaborating in real time through a combination of ICT and physical systems while protecting technologies against cyberattacks
Virtual reality	Simplifying, using visualisation to make decisions, and optimising using real-time data to mirror the physical world
Additive manufacturing	Process of creating 3D objects by layering material, based on digital models
Advanced materials	Creation of innovative substances with enhanced and unique performance characteristics for specialised applications

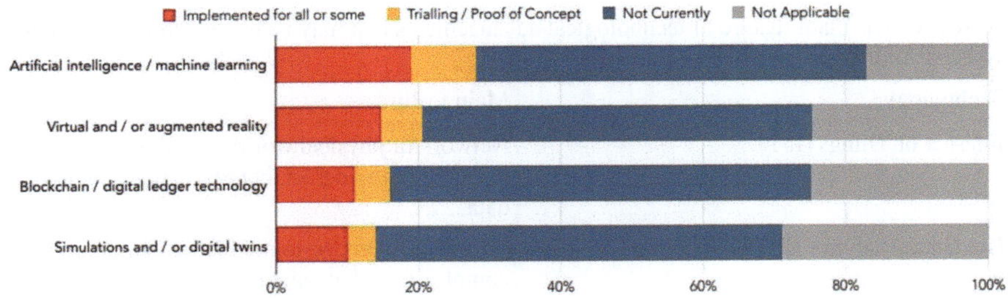

Fig. 8.1 Implementation of emerging technology initiatives among accounting professionals worldwide. *Source* ACCA (Association of Chartered Certified Accountants) 2023

provide more meaningful insights (Deniswara et al., 2023; Moll & Yigitbasioglu, 2019).

Prior literature examining the adoption of computer-assisted audit tools and techniques (CAATTs) among accounting professionals in various Asian contexts, such as Malaysia, Indonesia, and Jordan, highlights significant constraints on its implementation (Mansour, 2016; Nasrah et al., 2023; Purnamasari et al., 2022; Siew et al., 2020). The adoption of CAATTs is shaped by multifaceted factors, among which include the backing of top management, the conducive environment within audit firms, the technological proficiency of employees, and the perceived performance benefits (Mansour, 2016; Nasrah et al., 2023; Purnamasari et al., 2022; Siew et al., 2020).

Collectively, these studies underscore the limited current uptake of CAATTs, yet they also emphasise the potential for expanded usage given appropriate organisational and individual support. Additionally, the disparity between perception and implementation of advanced technology among accountants and auditors can stem from various factors such as considerations of return on investment, technical expertise, skill sets, and the complexities of determining the optimal applications of the technology in question. These discrepancies, often referred to as the perception-implementation gap, encapsulate the hurdles encountered in embracing and executing technological advancements, despite ongoing debates regarding their value and potential effectiveness (ACCA, 2023).

This chapter addresses the limitations of prior research by adopting and extending the Technology-Organization-Environment (TOE) framework (Tornatzky et al., 1990) and applying it to the audit domain. The chapter is structured as follows. The next section outlines the technological, organisational, and environmental factors influencing CAATTs adoption and introduces the Advanced CAATTs Adoption Typology. Following this, the chapter explores the psychological dimension of CAATTs adoption, recognising its crucial role in technology uptake. Next, it highlights the importance of overcoming existing barriers, fostering organisational support and motivation, encouraging auditor engagement, and tailoring adoption strategies to regional contexts. This is then followed by suggestions on steps that are critical for advanced CAATTs adoption. The chapter further underscores the need for future research to refine and empirically validate the proposed typology, investigate the integration of specific technologies in audit processes, incorporate human dimensions such as psychological factors, and examine the impact of cultural and institutional environments. Finally, the chapter concludes by emphasising the need for a holistic approach to advanced CAATTs adoption in the future, integrating technological and psychological dimensions through the TOE framework, and advocating for proactive engagement with emerging technologies to enhance audit efficiency, quality, and value in the evolving auditing landscape of Asia.

8.2 Key Concepts and Issues

8.2.1 Technological, Organisational and Environmental (TOE) Factors Influencing CAATTs (Computer-Assisted Audit Tools and Techniques) Adoption

This section advances the discourse on the adoption of advanced CAATTs by focusing on the pivotal influence of existing and emerging technologies, internal organisational resource capabilities, and external industry dynamics. In this context, the Advanced CAATTs Adoption Typology is developed using dimensions derived from the TOE Framework (technology-organisation-environment framework), which examines how new ideas, technologies, or practices spread within a social system over time (Jaakkola, 2020; Tornatzky et al., 1990). Originated from the Diffusion of Innovation (DOI) theory (Rogers Everett, 1995), the TOE Framework has undergone continual modification by prior research, leading to its broader use and increased theoretical validity (Alamgir Hossain & Quaddus, 2011; Awa et al., 2017).

8.2.1.1 Technological Context

In the context of CAATTs adoption, technology encompasses both the software and hardware employed in the auditing process. Adopters typically evaluate the efficacy of innovation based on its potential benefits and impediments. Pedrosa et al. (2019) noted that for statutory auditors in a European country, adopting technologies in auditing is crucial for enhancing efficiency and the effectiveness of their work. This adoption streamlines audit processes, reduces manual errors, and provides deeper insights through advanced data analytics, ultimately leading to more reliable and timely audit outcomes. These benefits may include enhanced service reliability, efficiency, and quality, while barriers may comprise factors such as complexity, compatibility issues, and the presence of legacy systems (Aviantib, 2019).

Moreover, the decision to adopt audit-related technology is often influenced by both subjective perceptions and organisational needs. This gives rise to the concept of task-fit, wherein organisations may opt against employing CAATTs if they do not align with their specific audit requirements, despite their recognised value as cutting-edge tools (Rosli et al., 2012). Currently, there exists limited understanding regarding the extent to which advanced CAATTs effectively fulfil audit work needs across diverse organisational contexts, particularly within audit firms of varying sizes. Additionally, the technological infrastructure and accessibility play pivotal roles in the adoption of CAATTs. The possession of robust infrastructure significantly influences the adoption and efficient utilisation of technology. However, further elucidation is warranted regarding CAATTs adoption in the Asian context (Aviantib, 2019).

Apart from that, the future of work in auditing, particularly in the Asian landscape, is being shaped by the integration of CAATTs and artificial intelligence (AI). These technologies are revolutionising traditional auditing practices, enhancing efficiency, accuracy, and risk assessment (Ganapathy, 2023; Odeyemi et al., 2023). In the public sector, the use of advanced CAATTs is being encouraged, with performance expectation and facilitating conditions playing a key role in their adoption (Purnamasari et al., 2022). The application of CAATTs in public accounting firms in Indonesia has been found to improve audit processes (Handoko et al., 2020). However, the adoption of AI-enabled auditing techniques in audit firms requires careful consideration of various factors, including data analysis and decision-making (Hu et al., 2020). The use of technology, such as the Medical Service Transformation and Innovation Compass (MYSTIC), has also been shown to enhance clinical audit in Asian emergency medical services (Ng et al., 2021). Despite these advancements, the unique features of various institutions continue to present opportunities for further research in the auditing field (Defond et al., 2020).

8.2.1.2 Organisational Context

The organisational context encompasses internal organisational attributes such as size, scope, structure, and communication methods, which may impact technology adoption (Tornatzky, Fleischer, & Chakrabarti, 1990). A supportive organisational environment that reinforces corporate values plays a crucial role in determining the adoption of CAATTs (Ahmi et al., 2014). In highly centralised organisations, senior management's decision to adopt CAATTs, despite opposition from lower levels, can facilitate its adoption (Troshani et al., 2011).

Additionally, organisational characteristics such as size, top management support and managerial encouragement are significant determinants of CAATTs adoption (Ahmi & Kent 2012; Curtis & Payne, 2008; Rosli et al., 2012). Larger firms are more inclined to adopt innovations due to financial incentives, whereas smaller organisations may exhibit greater adaptability due to their flexibility and concerns regarding adaptation (Troshani et al., 2011). Larger firms benefit from economies of scale and resilience to bear risks (Zhu et al., 2004), along with possessing greater financial and human resources to ensure successful technology utilisation (Bierstaker et al., 2014). Notably, the utilisation of CAATTs is often contingent upon firm size, with Big-4 firms emerging as pioneers in deploying these advanced tools (Ghani, 2016).

However, more recent studies indicate that other audit firms, including non-Big-4 entities, have also begun adopting CAATTs (Abdul Ghani et al., 2012; Deniswara et al., 2023).

Within the organisational perspective of the TOE framework, the human personnel context, specifically individual differences in attitude, perspectives, and motivation, influences technology adoption. Focusing on CAATTs adoption at the organisational level is advantageous as it involves the entire company rather than individual auditors. Moreover, Awuah et al. (2022) emphasised that firms should invest more in training workshops on big data analytics. Such training can enhance employees' skills and improve overall organisational performance. Nevertheless, organisational strategic focus and innovation adoption are highly influenced by individual peculiarities, with positive attitudes leading to motivation for technology adoption. Yet, there is a paucity of research on the impact of human variations on organisational focus and innovation adoption, particularly concerning the strategic focus on CAATTs adoption (Awa et al., 2015). According to Awa et al. (2015), the adoption of CAATTs at the organisational level is influenced by a range of factors, including individual differences in attitude, perspectives, and motivation. These factors are further shaped by personal characteristics and organisational context, such as risk propensity, self-efficacy, and resource availability (Tabak & Barr, 1999). In the specific context of cloud computing, the relative advantage, compatibility, management support, technological readiness, and competition pressure are key factors influencing adoption (Alkandi, 2022).

8.2.1.3 Environmental Context

The environmental context encompasses the overarching conditions within which an organisation operates, including competitors, industry dynamics, governmental regulations, and external influences (Tornatzky et al., 1990; Widuri et al., 2016). These factors play a crucial role in fostering technology adoption, as successful adoptions can serve as models for others to follow (Troshani et al., 2011). Sustainable organisations adeptly manage their external environments, which encompass market and industry structures, competitors, governmental regulations, and external pressures (Tornatzky et al., 1990).

The auditing of clients' management and operational systems is vital to ensure the accuracy of business activities and the proper generation of reports. As client size increases and systems become more complex, with larger volumes of data, the utilisation of superior technology, such as advanced CAATTs, by audit firms becomes increasingly imperative (Rosli et al., 2012). Furthermore, for CAATTs, the complexity of client's accounting information systems (AIS), professional body support, firm size, top management commitment, and employee IT competency are significant factors (Rosli, 2017; Rosli et al., 2012; Siew et al., 2020). The impact of Industry 4.0 technology adoption is determined by compatibility, top management support, and competitive pressures, with employee capability being more influential in the service industry (Zhong & Moon, 2023).

However, there remains a gap in understanding how specific environmental dynamics, such as societal external factors, uniquely influence technology adoption, particularly in the context of the audit field. Additionally, the impact of regulatory and professional bodies, especially in countries like Malaysia, on technology adoption strategies within audit firms is not thoroughly explored.

8.2.2 Advanced CAATTs (Computer-Assisted Audit Tools and Techniques) Adoption Typology

The TOE Framework facilitates a more comprehensive exploration of current CAATTs adoption. By identifying the elements and dynamics within the technology, organisation, and environment components, the TOE Framework elucidates the factors influencing its utilisation. Drawing from prior literature and the TOE Framework, this chapter presents a typology of the determinants of advanced CAATTs adoption (see Fig. 8.2), considering the technological advancements in audit tools and techniques in line with changes brought about by the Fourth Industrial Revolution (IR 4.0). The proposed Advanced CAATTs Adoption Typology aims to assist both individuals and organisations in keeping pace with contemporary developments by emphasising the critical factors from each of the technology, organisation, and environment components.

8.2.3 Human Aspects in CAATTs (Computer-Assisted Audit Tools and Techniques) Adoption: Strategies for Overcoming Psychological Challenges in Audit Firms

This section discusses the need to delve into the areas that demand careful consideration from auditors and audit firms in effectively adopting advanced CAATTs within the Asian context, including psychological aspects. The lack of adoption or barriers hindering the advancement of CAATTs can be multifaceted. While technological barriers such as compatibility issues, infrastructure requirements, and costs are often discussed, the psychological aspect is equally crucial but often underestimated.

Human beings are at the heart of the technology adoption processes, and their attitudes, perceptions, and behaviours significantly influence the success or failure of technology implementation. This involves acknowledging cultural attitudes towards technology, potential resistance to change, and the need for effective communication and training to overcome psychological barriers to adoption. Asian countries exhibit diverse cultural backgrounds, each with its own set of beliefs, values, and norms regarding the adoption of new technologies. Understanding and respecting these

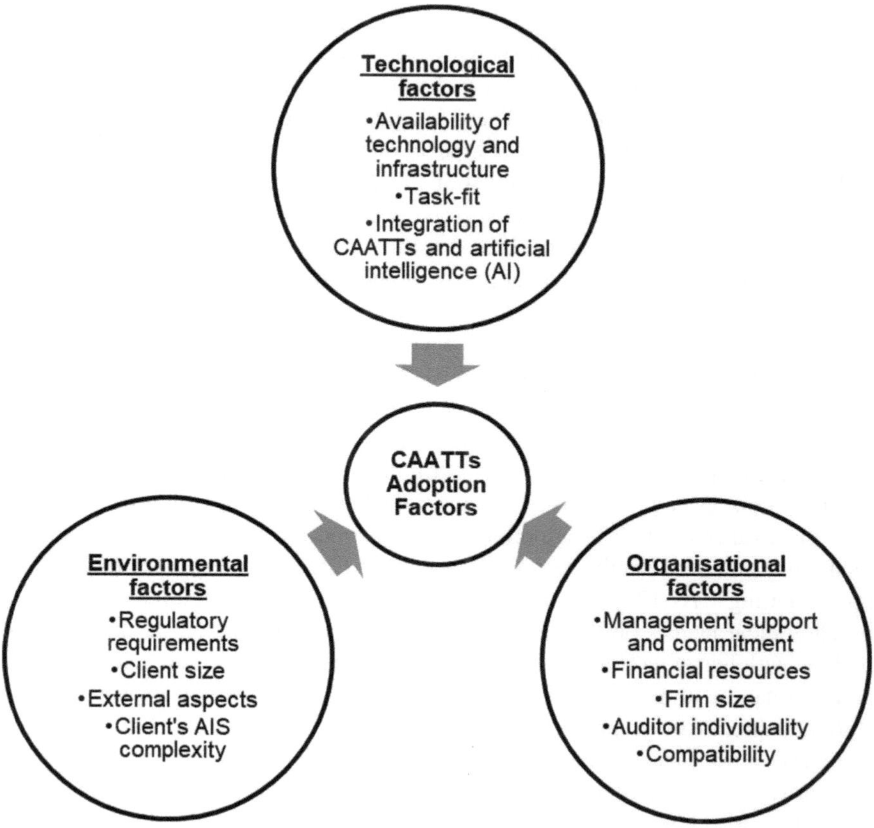

Fig. 8.2 Typology of advanced CAATTs (computer-assisted audit tools and techniques) adoption among auditors

cultural nuances are essential for designing effective adoption strategies that resonate with local stakeholders.

Additionally, it is crucial to address the potential resistance to change. Human beings, regardless of cultural background, tend to exhibit a degree of apprehension or reluctance towards unfamiliar technological innovations. This resistance may stem from a fear of job displacement, uncertainty about new processes, or simply a preference for familiar routines. Effective communication and training programmes tailored to address these concerns can help mitigate resistance and foster a culture of openness and receptiveness towards CAATTs.

Numerous studies in the field of technology adoption highlight the importance of considering the human aspect alongside the technical aspects. For instance, Kuhlmann (2014) highlighted the role of anxiety in shaping attitudes towards technology, while Vecchiotti and Small (1984) emphasised the need for a balanced approach that considers human factors. Reddy (1979) further underscored the importance of generating appropriate technologies that are aligned with user needs and capabilities. This is particularly relevant in the context of health technology, as discussed by Hastall et al. (2017), who calls for a user-centred framework that accounts for the process character of technology adoption, human motivation, and situational and social influences. These insights collectively suggest that a deeper understanding of the psychological aspects of technology adoption is crucial for addressing barriers to the advancement of CAATTs.

Additionally, understanding the specific needs and preferences of stakeholders in Asian countries can inform the customisation of CAATTs implementation strategies, ensuring alignment with local practices and expectations. By addressing these psychological dimensions alongside technical considerations, auditors can enhance the successful integration of advanced CAATTs in Asian audit practices. This holistic approach ensures that technological advancements are not only technically sound but also resonate with the human element, ultimately driving greater acceptance, effectiveness, and value in audit practices across the region.

In the fast-paced environment of audit firms, where auditors are constantly juggling multiple engagements and facing tight deadlines, the psychological component of advanced CAATTs adoption becomes even more pronounced. While the focus may often be on the technical aspects and financial resources required for implementing CAATTs, it is essential not to overlook the significant impact of psychological factors on the adoption process. As auditors navigate through busy periods with heavy workloads, the pressure to quickly adapt to new tools and technologies can be immense. This pressure is especially challenging for new auditors who may already be grappling with the demands of their roles.

Learning about advanced CAATTs amidst such demanding circumstances can be overwhelming, leading to feelings of stress and exhaustion. The complexity of CAATTs further compounds this challenge, as auditors may struggle to process and assimilate the influx of information, ultimately hindering their ability to effectively utilise these tools (Rosli et al., 2012). Moreover, the rapid evolution of technology adds another layer of complexity, contributing to feelings of anxiety and reluctance among auditors to change. Keeping pace with the continuous advancements in technology requires constant learning and adaptation, which can be daunting for some auditors. This anxiety is exacerbated by a fear of falling behind or being unable to meet the evolving expectations of their roles.

The stressful work environment prevalent in audit firms can exacerbate these psychological challenges. Auditors may perceive increased pressure as a barrier to embracing new technologies, leading to resistance and reluctance to change. This resistance is not only detrimental to individual well-being but also impedes the overall adoption of advanced CAATTs within the organisation. Despite these challenges, auditors' intrinsic motivation plays a crucial role in driving the adoption of CAATTs (Pedrosa & Costa, 2014; Searcy et al., 2003).

However, it is essential to recognise that diverse auditors have different learning styles and progress at different rates in terms of technology adoption. Tailoring training programmes and support mechanisms to accommodate these individual differences is key to fostering a more inclusive and effective adoption process (Nasrah et al., 2023). While the technical aspects of CAATTs adoption are undoubtedly important, it is equally critical to address the psychological factors that impact auditors' attitudes, beliefs, and emotions towards technology. By acknowledging and mitigating these psychological barriers, audit firms can create a more supportive and conducive environment for successful advanced CAATTs adoption, ultimately driving greater efficiency and effectiveness in audit practices.

The audit firms, especially the Big 4, play a pivotal role in facilitating the successful adoption of advanced computer-assisted audit tools and techniques (CAATTs) by fostering effective communication and providing comprehensive training programs. As highlighted by Searcy et al. (2003) and Pedrosa and Costa (2014), the audit firms can leverage their resources to implement training initiatives that equip their staff with the necessary skills and knowledge to effectively utilise CAATTs in audit processes. Comprehensive training programmes are essential for ensuring that auditors understand the functionalities and benefits of CAATTs. By offering clear and comprehensive information about the advantages of CAATTs, the audit firms can effectively communicate the value proposition of these tools to their auditors. Training sessions should cover various aspects of CAATTs, including their capabilities, usage scenarios, and potential impact on audit efficiency and effectiveness. Moreover, hands-on training and practical exercises can help auditors develop proficiency in using CAATTs, enabling them to confidently apply these tools in their day-to-day work.

In addition to training programmes, audit firms can provide motivation and support to auditors through robust moral encouragement and an unwavering commitment to change management initiatives. As emphasised by Molino et al. (2020), employee motivation is a key driver of technology adoption, and management support plays a crucial role in fostering a positive attitude towards change. By demonstrating a genuine commitment to the adoption of CAATTs and providing encouragement and recognition for auditors' efforts, management can instil a sense of ownership and enthusiasm among staff members.

Furthermore, offering opportunities for professional growth and skill development using technology can enhance employee motivation and engagement. Auditors are more likely to embrace CAATTs when they perceive the acquisition of new technological skills as a means to achieve both personal and professional development. Audit firms can leverage CAATTs as a platform for continuous learning and improvement, providing auditors with opportunities to expand their skill sets and stay abreast of emerging trends in audit technology. Regular feedback on progress and a culture of continuous improvement can reinforce employee participation in the adoption of CAATTs. By soliciting input from auditors and incorporating their feedback into the development and refinement of CAATT implementation strategies, audit firms can create an inclusive and collaborative environment that empowers auditors to take ownership of the adoption process.

In a nutshell, the audit firms, including non-Big 4 firms, play a crucial role in facilitating the adoption of advanced CAATTs by providing comprehensive training programmes, motivation, and support to auditors. By emphasising the benefits of CAATTs, fostering a positive attitude towards change, and offering opportunities for professional growth and skill development, these firms can create a conducive environment for successful technology adoption. Furthermore, audit firms can effectively integrate CAATTs into audit practices and drive greater efficiency and effectiveness by actively involving auditors in the adoption process and soliciting feedback.

8.3 Discussion of Implications

The integration of advanced CAATTs into audit practices presents a significant opportunity to enhance these practices. However, successful CAATTs adoption requires careful consideration of both technical and psychological aspects to overcome barriers and fully optimise the benefits of advanced technology. The Advanced CAATTs Adoption Typology offers a detailed understanding of each component and its respective elements, extending the existing literature by explicitly identifying current factors that support CAATTs adoption. By demonstrating how technological, organisational, and environmental components and their dynamics can act as enablers for CAATTs adoption, this chapter delves into psychological aspects such as cultural nuances, resistance to change, stakeholder diversity, and change adaptation. It provides guidance for advanced CAATTs adoption in Asia and in the global context.

It is vital to focus on specific factors that are fundamental to the future adoption of CAATTs. Overall, the conceptual ideas emphasise the necessity for audit firms to address the following building blocks to enhance audit efficiency through CAATTs: overcoming existing barriers, fostering organisational support and motivation, encouraging auditor engagement, and customising approaches to suit local contexts (see Fig. 8.3).

Technical barriers, such as compatibility issues and costs, along with psychological factors, such as resistance to change, digital fatigue, and stress from the rapid pace of work, hinder the adoption of CAATTs. Recognising and addressing these psychological dimensions are essential for creating a supportive environment conducive to successful technology adoption.

Auditors are often faced with high workloads and pressure, thus effective communication, comprehensive training programmes, and robust moral encouragement are crucial for facilitating the adoption of CAATTs. By providing adequate support and motivation, management can instil a sense of ownership and enthusiasm among auditors, thereby enhancing their willingness to embrace new technologies and acquire the necessary skills.

Furthermore, understanding the specific needs and preferences of stakeholders in Asian countries is essential for customising CAATTs implementation strategies and ensuring alignment with local practices and expectations. By tailoring adoption approaches to accommodate cultural attitudes towards technology and individual learning styles, auditors can foster a more inclusive and effective adoption process.

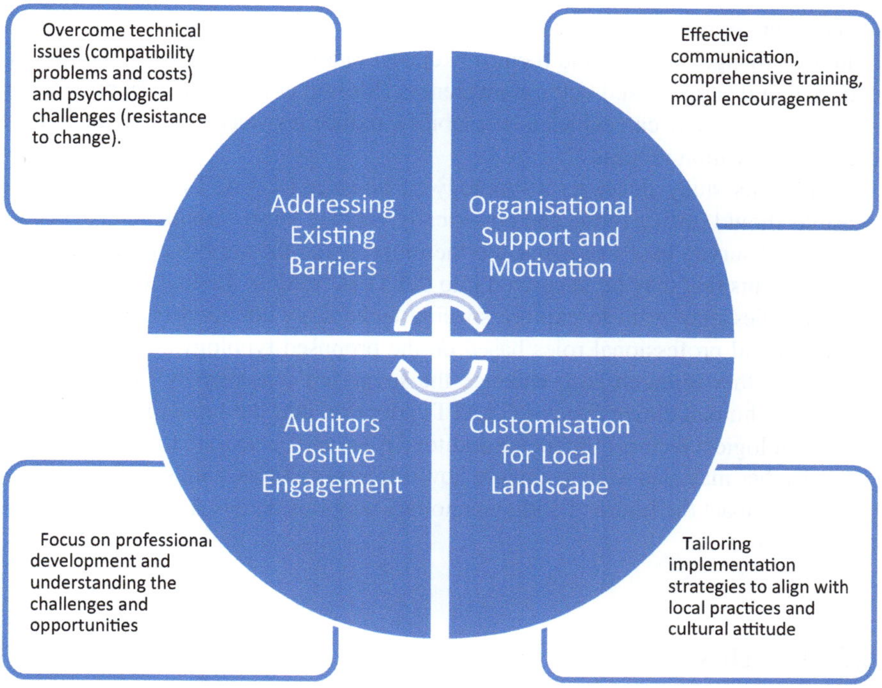

Fig. 8.3 Building blocks for audit efficiency with advanced CAATTs (computer-assisted audit tools and techniques) adoption

8.4 Future Directions

Looking ahead, our review suggests that audit firms take several key steps to successfully integrate advanced technologies such as AI, data analytics, and machine learning into their operations. First, they must conduct a thorough cost–benefit analysis to assess the value and feasibility of these investments. More importantly, the adoption of new technologies should be accompanied by efforts to optimise their functionality and ensure they align with the firm's specific audit processes (Hu et al., 2020). In other words, there must be a strong synergy between the firm's existing systems and the newly implemented technology (Handoko et al., 2020; Ng et al., 2021; Pedrosa et al. 2019). While adopting these tools is a critical step, the real challenge lies in effectively utilising them to enhance efficiency and maintain high audit standards.

Another notable factor audit firms must address is the readiness of their workforce to adapt to new technologies. As technology continues to evolve, auditors need to upskill to stay efficient and competitive in a rapidly changing landscape. To ensure a smooth transition, firms should invest in comprehensive training programmes, skill development initiatives, and robust infrastructure (Awuah et al., 2022). The audit firms must strategically address the psychological aspects of technology adoption, recognising the diverse backgrounds, generational differences, and unique strengths of their human resource (Nasrah et al., 2023). To foster a positive outlook on technological change, firms should focus on key soft factors such as effective communication, moral support, and cultural awareness (Molino et al., 2020). By prioritising these elements, they can enhance adaptability, reduce resistance, and create a more inclusive transition process.

While this study develops a typology of advanced CAATTs adoption, future research should further refine and empirically validate these classifications via case studies and survey-based research. Furthermore, given the variety of technological advancements that may be integrated into CAATTs (refer to Table 8.1 in Sect. 8.1), future studies may want to explore specific technology adoption in specific audit processes and professional roles based on the proposed typology. Another critical area for further research is to enhance the suggested typology by explicitly integrating the human dimensions of CAATTs adoption into the typology, particularly the psychological factors influencing auditors' roles and behaviour. Last but not least, future studies may also want to explore how different cultures, and institutional environments impact the future of audit technology adoption across Asia and other parts of the world.

8.5 Conclusion

This chapter underscores the importance of further research into the utilisation of advanced technology to promote audit efficiency and quality by applying the TOE framework to identify factors influencing CAATTs adoption. The proposed advanced CAATTs' Adoption Typology identifies the technological, environmental, and organisational factors that enable CAATTs adoption. In line with the rise of web-based technology in supporting audit practices, this typology highlights several contemporary factors, such as the integration of artificial intelligence, the complexity of clients' accounting information systems, and the auditor's individuality and compatibility in handling new technology. These new areas are among the essential focal points for audit firms.

Second, by incorporating the psychological aspects of CAATTs adoption, this chapter critically integrates both soft and hard factors into its discussion. It cohesively considers human and technical factors to optimise CAATTs adoption to support auditors' work. Finally, the chapter outlines the building blocks of CAATTs adoption, emphasising proactive engagement with technological advancements, ongoing professional development, and a thorough understanding of the unique challenges and opportunities presented by CAATTs' adoption. By prioritising these areas, auditors in Asia and globally can leverage advanced CAATTs to enhance audit quality and deliver greater value to stakeholders in the evolving landscape of auditing practices.

In summary, the successful integration of advanced CAATTs into audit practices requires a holistic approach that considers both technical and psychological dimensions. By addressing barriers and leveraging opportunities, auditors and audit firms can harness the full potential of advanced technology to drive innovation, improve efficiency, and uphold audit quality in an ever-changing business environment.

Highlights

- Technological transformation in audit work—The rise of advanced CAATTs integrating technologies such as AI, machine learning, big data analytics and cloud-based tools.
- New skills requirements for auditors—Auditors need to be tech-savvy and agile to leverage CAATTs.
- Hybrid work models—More integration between humans and technology soon, requiring both technical and psychological support in audit firms.
- Overcoming skill gaps and resistance to change—Successful implementation of CAATTs requires effective communication that acknowledges the different strengths and weaknesses of auditors from various age groups and environments.
- The future of audit in Asia—The need for audit firms in Asia to strategically manage opportunities and challenges of advanced CAATTs.

References

Abdul Ghani, A., Shahimi, S., & Che Azmi, A. (2022). Determinants of computer assisted audit tools and techniques (CAATs) adoption. *Advanced International Journal of Banking Accounting and Finance, 4*(12), 01–21.

ACCA. (2023). *Digital horizons: Technology, innovation and the future of accounting*. Retrieved April 20, 2024, from https://www.accaglobal.com/content/dam/ACCA_Global/professional-insights/digital-horizons2023/PI-DIGITAL-HORIZONS%20v5.pdf

Ahmi, A., & Kent, S. (2012). The utilisation of generalized audit software (GAS) by external auditors. *Managerial Auditing Journal, 28*(2), 88–113. https://doi.org/10.1108/02686901311284522

Ahmi, A., Saidin, S. Z., & Abdullah, A. (2014). IT adoption by internal auditors in public sector: A conceptual study. *Procedia-Social and Behavioral Sciences, 164*, 591–599. https://doi.org/10.1016/j.sbspro.2014.11.151

AICPA. (2023). *DAS: The future of audit is now*. Retrieved April 21, 2024, from https://www.cpa.com/sites/cpa/files/media/resources/whitepapers/DAS-The-future-of-audit-is-now.pdf

Alamgir Hossain, M., & Quaddus, M. (2011). The adoption and continued usage intention of RFID: An integrated framework. Edited by J. R. Venable. *Information Technology & People, 24*(3), 236–256. https://doi.org/10.1108/09593841111158365

Alkandi, I. (2022). Technological, organizational, and environmental factors influencing the adopting of cloud computing: A quantitative study. In *12th International scientific conference business and management 2022*. https://doi.org/10.3846/bm.2022.692

Aviantib, S. M. I. (2019). Improving audit quality: Adopting technology and risk management. *International Journal of Innovation, Creativity and Change, 8*(3), 89–103.

Awa, H. O., Ojiabo, O. U., & Emecheta, B. C. (2015). Integrating TAM, TPB and TOE frameworks and expanding their characteristic constructs for E-commerce adoption by SMEs. *Journal of Science and Technology Policy Management, 6*(1), 76–94. https://doi.org/10.1108/jstpm-04-2014-0012
Awa, H. O., Ojiabo, O. U., & Orokor, L. E. (2017). Integrated Technology-Organization-Environment (T-O-E) taxonomies for technology adoption. *Journal of Enterprise Information Management, 30*(6), 893–921. https://doi.org/10.1108/JEIM-03-2016-0079
Awuah, B., Onumah, J. M., & Duho, K. C. T. (2022). Determinants of adoption of computer-assisted audit tools and techniques among internal audit units in Ghana. *The Electronic Journal of Information Systems in Developing Countries, 88*(2), Article e12203. https://doi.org/10.1002/isd2.12203
Bierstaker, J., Janvrin, D., & Lowe, D. J. (2014). What factors influence auditors' use of Computer-Assisted Audit techniques? *Advances in Accounting, 30*(1), 67–74. https://doi.org/10.1016/j.adiac.2013.12.005
Curtis, M. B., & Payne, A. E. (2008). An examination of contextual factors and individual characteristics affecting technology implementation decisions in auditing. *International Journal of Accounting Information Systems, 9*(2), 104–121. https://doi.org/10.1016/j.accinf.2007.10.002
Defond, M. L., Zhang, F., & Zhang, J. (2020). Auditing research using Chinese data: What's next? *Accounting and Business Research*, April, 1–14. https://doi.org/10.1080/00014788.2020.1746626
Deniswara, K., Henky, T., Mulyawan, A. N., Armand, W. K., & Mustapha, M. (2023). The role of external auditor in the adoption of computer assisted audit techniques (CAATS) with unified theory of acceptance and use of technology: An empirical study in Public Audit firms in Jakarta. *The Winners, 24*(1), 1–11. https://doi.org/10.21512/tw.v24i1.8124
Ganapathy, V. (2023). AI in auditing: A comprehensive review of applications, benefits and challenges. *Shodh Sari-an International Multidisciplinary Journal, 02*(04), 328–343. https://doi.org/10.59231/sari7643
Ghani, R. (2016). Adoption of computer-assisted audit tools and techniques (CAATTs): An exploratory study in audit firms. In *International conference on accounting studies (ICAS) 2016.*
Handoko, B. L., Lindawati, A. S., & Mustapha, M. (2020). Application of computer assisted audit techniques in public accounting firm. Rochester, NY. https://papers.ssrn.com/sol3/papers.cfm?abstract_id=3629047
Handoyo, S. (2024). Mapping the landscape of internal auditing effectiveness study: A bibliometric approach. *Cogent Business & Management, 11*(1). https://doi.org/10.1080/23311975.2023.2289200
Hastall, M. R., Dockweiler, C., & Mühlhaus, J. (2017). Achieving end user acceptance: Building blocks for an evidence-based user-centered framework for health technology development and assessment. In *Universal access in human–computer interaction. Human and technological environments* (pp. 13–25). https://doi.org/10.1007/978-3-319-58700-4_2
Hu, K.-H., Chen, F.-H., Hsu, M.-F., & Tzeng, G.-H. (2020). Identifying key factors for adopting artificial intelligence-enabled auditing techniques by joint utilization of fuzzy-rough set theory and MRDM technique. *Technological and Economic Development of Economy, 27*(2), 459–492. https://doi.org/10.3846/tede.2020.13181
Jaakkola, E. (2020). Designing conceptual articles: Four approaches. *AMS Review, 10*(1–2), 18–26. https://doi.org/10.1007/s13162-020-00161-0
Kuhlmann, A. (2014). Problems associated with the acceptance of new technologies in industrialized societies. *International Journal of Technology Management*. https://www.semanticscholar.org/paper/Problems-associated-with-the-acceptance-of-new-in-Kuhlmann/27570ca6ca987636569cf80207755f39e7557b79
Liew, A., Boxall, P., & Setiawan, D. (2022). The transformation to data analytics in big-four financial audit: What, why and how? *Pacific Accounting Review, 34*(4), 569–584. https://doi.org/10.1108/PAR-06-2021-0105

Mansour, E. M. (2016). Factors affecting the adoption of computer assisted audit techniques in audit process: Findings from Jordan. *Business and Economic Research Macrothink Institute, 6*(1), 248–271.

Molino, M., Cortese, C. G., & Ghislieri, C. (2020). The promotion of technology acceptance and work engagement in Industry 4.0: From personal resources to information and training. *International Journal of Environmental Research and Public Health, 17*(7), 2438. https://doi.org/10.3390/ijerph17072438

Moll, J., & Yigitbasioglu, O. (2019). The role of internet-related technologies in shaping the work of accountants: New directions for accounting research. *The British Accounting Review, 51*(6), Article 100833. https://doi.org/10.1016/j.bar.2019.04.002

Nasrah, H., Muda, I., & Kesuma, S. (2023). Computer assisted audit tools and techniques adoption: A systematic literature review. *International Journal of Social Service and Research, 3*(3), 630–638.

Ng, Q. X., Yeung, W. L. K., Tay, J. A. M., & Arulanandam, S. (2021). Use of technology to aid clinical audit in an Asian emergency medical services department". *Healthcare, 9*(5), 491. https://doi.org/10.3390/healthcare9050491

Odeyemi, O., Awonuga, K. F., Mhlongo, N. Z., Ndubuisi, N. L., Olatoye, F. O., & Daraojimba, A. I. (2023). The role of AI in transforming auditing practices: A global perspective review. *World Journal of Advanced Research and Reviews, 21*(2), 359–370. https://doi.org/10.30574/wjarr.2024.21.2.0460

Pedrosa, I., & Costa, C. J. (2014). New trends on CAATTs. *Proceedings of the international conference on information systems and design of communication—ISDOC '14*. https://doi.org/10.1145/2618168.2618190

Pedrosa, I, Costa, C. J., & Aparicio, M. (2019). Determinants of adoption of Computer-Assisted Auditing Tools (CAATs). *Cognition, Technology & Work*, July. https://doi.org/10.1007/s10111-019-00581-4

Purnamasari, P., Amran, N. A., & Hartanto, R. (2022). Modelling computer assisted audit techniques (CAATs) in enhancing the Indonesian public sector. *F1000Research, 11*, 559. https://doi.org/10.12688/f1000research.121674.1

Reddy, A. K. (1979). *Problems in the generation of appropriate technologies* (pp. 173–189). Palgrave Macmillan UK EBooks. https://doi.org/10.1007/978-1-349-03931-9_8

Rogers Everett, M. (1995). *Diffusion of innovations*. New York, 12.

Rosli, K. (2017). Investigating the technological, organisational and environmental influences on the adoption of Computer-assisted Audit Tools and Techniques among audit firms in Malaysia. Monash University. https://doi.org/10.4225/03/58af74f4aaa02

Rosli, K., Yeow, P. H. P., & Eu-Gene, S. (2012). factors influencing audit technology acceptance by audit firms: A new i-TOE adoption framework. *Journal of Accounting and Auditing: Research & Practice,* 1–11. https://doi.org/10.5171/2012.876814

Searcy, D., Woodroof, J., & Behn, B. (2003). Continuous audit: The motivations, benefits, problems, and challenges identified by partners of a Big 4 accounting firm. In *36th annual Hawaii international conference on system sciences*. https://doi.org/10.1109/hicss.2003.1174565

Siew, E.-G., Rosli, K., & Yeow, P. H. P. (2020). Organizational and environmental influences in the adoption of computer-assisted audit tools and techniques (CAATTs) by audit firms in Malaysia. *International Journal of Accounting Information Systems*, 100445. https://doi.org/10.1016/j.accinf.2019.100445

Tabak, F., & Barr, S. H. (1999). Propensity to adopt technological innovations: The impact of personal characteristics and organizational context. *Journal of Engineering and Technology Management, 16*(3–4), 247–270. https://doi.org/10.1016/s0923-4748(99)00011-9

Tornatzky, L. G., Fleischer, M., & Chakrabarti, A. K. (1990). *The processes of technological innovation*. Lexington Books.

Troshani, I., Jerram, C., & Rao Hill, S. (2011). Exploring the public sector adoption of HRIS. *Industrial Management & Data Systems, 111*(3), 470–488. https://doi.org/10.1108/02635571111118314

Vecchiotti, R. A., & Small, A. M. (1984). Human factors and technology: The user sets the pace. *Springer EBooks* (pp. 295–300). https://doi.org/10.1007/978-1-4613-2401-0_28

Widuri, R., O'Connell, B., & Yapa, P. W. S. (2016). Adopting generalized audit software: An Indonesian perspective. *Managerial Auditing Journal, 31*(8/9), 821–847. https://doi.org/10.1108/MAJ-10-2015-1247

Zhong, Y., & Moon, H. C. (2023). Investigating the impact of industry 4.0 technology through a TOE-based innovation model. *Systems, 11*(6), 277. https://doi.org/10.3390/systems11060277

Zhu, K., Kraemer, K. L., & Dedrick, J. (2004). Information technology payoff in e-business environments: An international perspective on value creation of e-business in the financial services industry. *Journal of Management Information Systems, 21*(1), 17–54. https://doi.org/10.1080/07421222.2004.11045797

Suhaily Shahimi is a senior lecturer at the Department of Accounting, Faculty of Business and Economics, Universiti Malaya. Professional in the field of education who holds a Doctor of Philosophy (PhD) degree from Universiti Malaya with a concentration in Internal Audit and Corporate Governance; with extensive experience and a documented history of working in the education management business; skilled in Internal Audit, Risk Management, Governance, Analytical Skills, Lecturing, and Research. The subjects taught included advanced auditing, internal auditing, accounting integrated case studies, MBA internal audit courses, and a Master of Accounting (Reporting and Management Accountability) research methodology course. The research areas include auditing, risk management, governance, ESG, data analytics, and artificial intelligence. Suhaily received the Best Presenter Award 2021 and Best Paper Award 2021 (4th International Conference on Accounting Research and Education).

Dayana Jalaludin is a senior lecturer at the Accounting Department, Faculty of Business and Economics, Universiti Malaya. Her teaching areas include management accounting (basic, intermediate, and high-level), integrated case studies, and social and environmental accounting. Her research focuses on behavioural accounting in management accounting, accounting education, auditing, corporate governance, and the public sector. Dayana has published in various journals including Social Responsibility Journal, Journal of Cleaner Production, Accounting Education and Sustainable Development.

Ravinder Kaur Harminder Singh is a Master of Accounting student at Universiti Malaya. She holds a Bachelor of Accounting with Honours from Universiti Utara Malaysia and was recognised for academic excellence with the Deputy Vice Chancellor (Academic and International) Award. Ravinder brings with her valuable experience as an Audit & Assurance Intern at Crowe Malaysia PLT.

Chapter 9
Future Digital Work Era: Well-Being, Technostress and Turnover Intention

Clusters of Future Studies: Past and Futures; Technological Trends

Phuong Ngoc-Duy Nguyen, Cong Hiep Duong, Thi Trang Giang, and Huan Hong Nguyen

Abstract This empirical chapter investigates the impact of new ways of working (NWW), enabled by technology, on employee turnover, focusing on the Asian workforce. The objective is to understand how technological advancements in work practices reshape present and future work environments and influence employee dynamics. Emphasising future-oriented implications, the chapter considers the Vietnamese workforce as a representative context within the region, anticipating how these dynamics may unfold in Asia's evolving digital economies. By analysing past findings from other Asian countries, this study not only reflects on current realities but also provides future-looking insights into workforce transformation. Employing structural equation modelling for analysis, the chapter uses a comprehensive dataset of Vietnamese employees from various cities. This approach facilitates exploring the complex relationships between NWW, employee well-being, technostress, and turnover intention. The findings reveal the dual nature of the NWW, demonstrating its capacity to both elevate technostress and enhance employee well-being. Distinctively diverging from earlier research, it reveals that NWW is insignificant regarding

P. N.-D. Nguyen
School of Economics, Finance and Accounting, International University, Vietnam National University, Ho Chi Minh City, Vietnam
e-mail: nndphuong@hcmiu.edu.vn

C. H. Duong · T. T. Giang
Faculty of Accounting, Thai Nguyen University of Economics and Business Administration, Thai Nguyen, Vietnam
e-mail: hiepdc@tueba.edu.vn

T. T. Giang
e-mail: giangtrang.tn@tueba.edu.vn

H. H. Nguyen (✉)
Computer Science Department, Maharishi International University, Iowa, USA
e-mail: huan.dhsp@gmail.com

C. Y. Fung et al. (eds.), *Future of Work in Asia*,
https://doi.org/10.1007/978-3-031-95804-5_9

employee turnover intention. The mechanism of action appears rooted in the escalation of technostress, yet it remains unaffected by its implications on employee well-being. This exploration contributes a nuanced understanding of NWW's complex effects within Vietnam's unique socio-economic landscape and the broader future of work in Asia.

Keywords Digital work · Technostress · Well-being · Turnover intention · New ways of working

9.1 Introduction

Rapid technological progress has dramatically transformed workplace dynamics, ushering in a new era of labour practises characterised by digital nomadism, flexible scheduling, and enhanced productivity. These transformations represent not only a shift in present conditions but also a signal of future work trajectory. This paradigm shift, known as the new ways of working (NWW), has granted employees remarkable autonomy, blurring traditional boundaries of workspace and worktime (Alfes et al., 2022; Gerards et al., 2018; Metselaar et al., 2023). However, this shift raises concerns about employee well-being (EWB), especially considering the mental health impacts of digitalisation (Campbell & Gavett, 2021; Chillakuri & Vanka, 2021). In the context of the future of work in Asia, understanding these impacts is critical for preparing resilient and sustainable workforce strategies.

Despite NWW's positive aspects, its potential drawbacks like increased stress, technostress (TNS), and turnover have become crucial considerations for the evolving future of work (Jang & Kandampully, 2018; Jensen, 2021; Maier et al., 2022; Yasin et al., 2023). The inconsistency in understanding the effects of NWW on turnover intention (TI) is notable, with TNS resulting from technology interactions—emerging as a key factor impacting employee commitment and turnover (Andrulli & Gerards, 2023; Bainbridge & Townsend, 2020; Caillier, 2018; Jaiswal et al., 2022; Tarafdar et al., 2010). This study recognises a gap in understanding NWW's adverse effects on employees, particularly in non-Western contexts like Vietnam, a rapidly growing economy with unique socio-economic factors (Akbari & Hopkins, 2022; Han, 2023; Le et al., 2022; Maier et al., 2022; Nastjuk et al., 2023; Zhu et al., 2013). The local cultural, economic, and political nuances of Vietnam, which significantly differ from Western norms, play a vital role in shaping employee behaviour and attitudes toward job retention or turnover (Alpar, 2020; Hofstede, 1993).

The research addresses two key questions: How do new working practises affect TI among Vietnamese employees? And what roles do EWB and TNS play as mediators between NWW and TI within the Vietnamese workforce? By exploring these aspects, the study aims to provide insights into managing and reducing turnover in non-Western contexts, contributing to both academic knowledge and practical strategies for enhancing employee retention and well-being in the face of technological changes and the anticipated future of work across Asia (Rubenstein et al., 2018). This research not only delves into the direct impact of NWW on employee TI in Vietnam but also investigates the psychological intermediaries of EWB and TNS, offering a comprehensive view of the changing workplace landscape in a developing country context.

The study contributes significantly to the NWW literature by investigating the intricate relationship between technological workplace transformations and their repercussions on employee TI within a non-Western context, specifically in Vietnam. This focus is particularly pertinent given the rapid socio-economic changes and the distinct cultural landscape of Vietnam, which diverges markedly from Western norms that predominantly dominate existing research. The investigation into how NWW influences TI through the mediatory roles of EWB and TNS enhances our understanding of the psychological and emotional challenges that accompany digital work practices. By contextualising the effects of NWW in Vietnam, the research addresses a critical gap in the current understanding of global work practices and their local impacts. This is crucial as the adaptability and reaction to NWW can significantly vary based on cultural, economic, and political environments. The study's exploration of EWB and TNS as mediating factors provides a nuanced analysis of the pathways through which NWW can lead to increased employee turnover, offering academically novel and practically relevant insights. Furthermore, situating the study within the Vietnamese workforce contributes to a more global discourse on labour practices, moving beyond the typical Western-centric perspectives. This broader approach facilitates a more inclusive understanding of how technological advancements in the workplace reshape employee dynamics worldwide. The findings are expected to inform strategies that can mitigate adverse outcomes such as high turnover rates and diminished EWB, thereby aiding organisations in crafting more effective and culturally attuned human resource policies that foster productivity and employee satisfaction in the digital age.

9.2 Literature Review

Grounded in the Conservation of Resources (COR) theory (Hobfoll, 1989) and the Job Demands-Resources (JD-R) theory (Demerouti et al., 2001), the examination of new ways of working (NWW) reveals complex interactions between technological integration, employee well-being (EWB), and work dynamics. The COR theory specifically highlights how NWW introduces technostress (TNS) as a significant stressor while enhancing efficiency and connectivity. This arises from constant connectivity and blurred work-life boundaries, leading to resource depletion and increased work-family conflicts (Andrulli & Gerards, 2023; Harris et al., 2022; Hobfoll, 2001; Suh & Lee, 2017). Concurrently, the JD-R theory offers insights into how NWW introduces beneficial job resources like flexible scheduling and improved access to organisational knowledge. These resources help mitigate the negative impacts of job demands, fostering improved work engagement, deepening colleague connectivity, and enhancing overall psychological well-being, thereby contributing to positive outcomes like increased job satisfaction and reduced turnover intention (TI) (Bakker et al., 2023; De Vincenzi et al., 2022; Gajendran & Harrison, 2007; Mahomed et al., 2023; Yadav, 2016). NWW, characterised by non-traditional, flexible work strategies, has significantly transformed modern employment practices. These strategies are designed to optimise productivity and harmonise work-life balance by incorporating advanced information and communication technologies. NWW represents not only a shift in tools used but also a paradigm shift in workplace culture, emphasising flexibility and outcome-focused management (Bellmann & Hübler, 2021; Gerards et al., 2021; Ollo-López et al., 2021). Key objectives of NWW include reducing overhead expenses, enhancing productivity, fostering knowledge exchange, improving employee satisfaction, and encouraging collaborative efforts (Angelici & Profeta, 2023). Accordingly, TI, or the likelihood of employees leaving their positions, is influenced by these NWW-induced changes. Recent studies have suggested that the flexibility afforded by remote working under NWW can promote work-life balance, thus lowering TI and enhancing employee loyalty (Bainbridge & Townsend, 2020; Jamal et al., 2023; Ketkaew et al., 2020; Puhakka et al., 2021; Shanmugam & Agarwal, 2019). Therefore, it is hypothesised:

H1: NWW has a negative impact on TI

The rise of technological advancements brings the latent threat of TNS. This form of stress intensified in NWW contexts, arising from the demands of being constantly connected and managing the overlap of professional and personal life. As the demarcation between occupational tasks and personal commitments becomes increasingly blurred, individuals grapple with intensifying pressures, striving to juggle professional obligations with familial responsibilities equitably (Ketkaew et al., 2020). Research indicates that this escalation of stress can lead to adverse psychological and behavioural outcomes, potentially affecting workers' intentions to stay with their employer (Burns et al., 2016; Hwang & Cha, 2018; Molino et al., 2020; Narahari & Koneru, 2017; Tarafdar et al., 2007). Thus, the second hypothesis is proposed:

H2: NWW significantly indirectly impacts employee TI through the mediating role of TNS.

NWW practises, such as flexible scheduling and enhanced access to organisational knowledge, can be conceptualised as job resources that bolster work engagement by fostering efficient communication and deepening connectivity among colleagues, subsequently improving overall psychological well-being (Bakker et al., 2023). Job autonomy, effective communication, and social support are vital factors that contribute positively to EWB (Gajendran & Harrison, 2007; Mahomed et al., 2023; Yadav, 2016). By granting individuals greater control over their surroundings and scheduling, NWW allows them to tailor their work to fit their unique needs and preferences, resulting in greater job satisfaction. According to Wang et al. (2021), employees who are pleased and in good psychological condition are more inclined to be dedicated and productive, making them less likely to contemplate leaving their organisations. In addition, Cao et al. (2023) note that when employees achieve a sense of well-being, their commitment to the organisation strengthens, reducing their intention of leaving. Therefore, the third hypothesis is formulated:

H3: NWW significantly indirectly impacts TI by mediating EWB.

Empirical evidence supports the idea that employees who enjoy psychological health and happiness fostered by NWW practices tend to be more dedicated and productive and, consequently, less likely to contemplate leaving their organisations. This positive impact is partly due to well-being initiatives inherent in NWW, such as flexible scheduling, which supports an appropriate balance between work and life and reduces the temptation to seek new job opportunities. Moreover, when employees experience a sense of fulfilment and well-being, their commitment to the organisation strengthens, reducing their likelihood of leaving. This notion is affirmed by the research of Wang et al. (2021), Haar and Brougham (2022), and Xiao et al. (2023). On the contrary, Suh and Lee (2017) highlight the negative impact of technology on employees, warning that the demand for "anytime, anywhere connectivity" can increase stress and effort, harming mental well-being and job satisfaction, potentially leading employees to consider other job options. This sentiment is echoed by Gillespie et al. (2001) and Narahari and Koneru (2017). Given these dynamics, the study proposes a complex relationship where NWW can positively and negatively affect TI, mediated by TNS and EWB. While NWW can potentially lead to increased TNS, adversely impacting TI, it also offers resources that enhance EWB, which can mitigate the likelihood of employees leaving their jobs. This complex interplay underscores the nuanced impact of modern work practices on employee retention and turnover. Thus, the final hypothesis is proposed:

H4: The relationship between NWW and TI is serially mediated by TNS and EWB.

The conceptual research model is presented in Fig. 9.1.

9.3 Methodology

This study utilises a quantitative, cross-sectional research methodology to explore professionals' perspectives on new ways of working (NWW), technostress (TNS), employee well-being (EWB), and turnover intention (TI) within the Vietnamese context (see Fig. 9.1). The aim is to understand the crucial determinants and their interconnections in contemporary occupational settings, mainly focusing on the significant influence these elements exert on each other within Vietnam's socio-cultural environment.

Initially in English, the questionnaire was translated into Vietnamese using the back-translation method (Brislin, 1976). Its clarity was verified through preliminary assessments involving a group of eight independent scholars and ten prospective participants. These evaluations led to refinements in the questionnaire, ensuring its comprehensibility and relevance. Accompanying each questionnaire was a cover

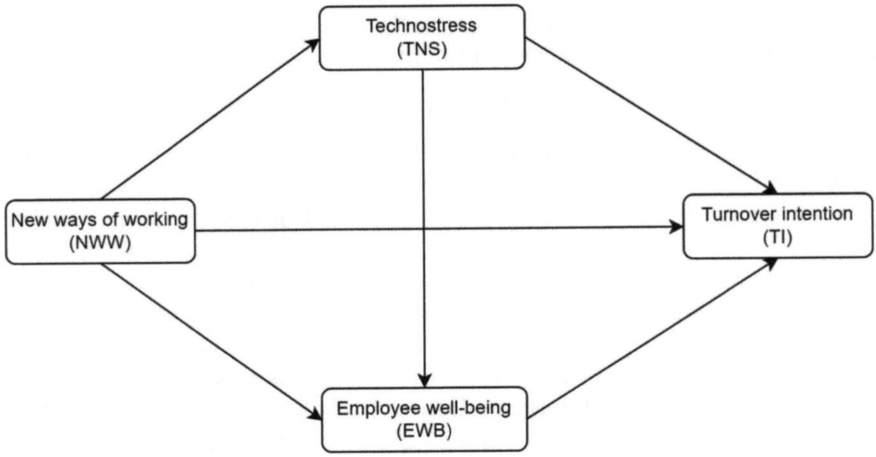

Fig. 9.1 Conceptual framework exploring how new ways of working (NWW) affects turnover intention (TI) through the mediation of technostress (TNS) and employee well-being (EWB)

letter detailing the research's purpose and significance, aiding participants in fully understanding the content. The questionnaire consisted of two parts: demographic information and variable measures. The demographic section collected data on age, gender, location, and occupation, while the variable measures were based on scales from prior studies. NWW was measured with ten items from Gerards et al. (2018), TNS with three items from Jaiswal et al. (2022), EWB with eight items also from Jaiswal et al. (2022), and TI through three items based on Dysvik and Kuvaas (2013). Responses were gathered using a 5-point Likert scale.

Data collection, conducted between March and June 2023, utilised Google Forms and targeted cities across Vietnam. Participants were reached through personal and professional networks, with invitations sent face-to-face, via email, and on social media platforms. In line with Hair et al. (2021), the sample size was determined to be at least ten times the most significant number of structural paths directed at a latent construct in the model. 336 validated responses were collected, providing a robust dataset for analysing the proposed constructs. Table 9.1 displayed comprehensive demographic details of the participants, emphasising the significance and effectiveness of the data collection process.

We used SmartPLS 3.2.9 for data analysis with a sample size of over 300 participants. Therefore, we selected PLS-SEM and SPSS 25 for descriptive analysis of demographic data. The measurement model was evaluated first for reliability (Cronbach's alpha, Composite Reliability), convergent validity (factor loadings and Average Variance Extracted, or AVE), and discriminant validity (Fornell and Larcker, HTMT). After the measurement model proved satisfactory, the structural model was examined, and 5000 resamples of the bootstrapping approach were used to evaluate hypotheses (Fornell & Larcker, 1981; Hair et al., 2010, 2019). This comprehensive

Table 9.1 Demographic characteristics of the survey participants in Vietnam

Characteristics	Distribution	Frequency	Percentage (%)
Gender	Male	113	33.63
	Female	223	66.37
Age	18–25	99	29.46
	26–35	148	44.05
	36–45	78	23.21
	46–55	9	2.68
	> 55	2	0.60
Education level	High school and below	28	8.33
	University/College	247	73.51
	Master and above	61	18.15
Marital status	Single	134	39.88
	Married	202	60.12

method ensures the validity of the results, providing an in-depth understanding of the relationships between NWW, TNS, EWB, and TI.

9.4 Findings

9.4.1 Measurement Model Analysis of the Conceptual Framework

Our research underwent rigorous scrutiny to ascertain the precision of the measurement model. As shown in Table 9.2, the reliability of the constructs was confirmed through Cronbach's Alpha indices and composite reliability, both of which surpassed the benchmark of 0.70 (Hair et al., 2021). Construct validity was evaluated bifurcated. The constructs exhibited convergent validity, as both factor loadings and AVE for each exceeded the recognised norms of 0.7 and 0.5 in that order (Hair et al., 2010).

Discriminant validity, emphasising the distinctiveness of each construct, was confirmed following the guidelines set by Fornell and Larcker (1981). This was further supported by Heterotrait-Monotrait Ratio (HTMT) values below 0.90, as indicated in Henseler et al. (2015)'s study, and presented in Tables 9.3 and 9.4.

Table 9.2 Result of reliability and convergent validity of constructs related to new ways of working (NWW), technostress (TNS), employee well-being (EWB), and turnover intention (TI)

Construct	Cronbach's Alpha	CR	AVE
NWW	0.908	0.926	0.611
TI	0.916	0.946	0.853
TNS	0.838	0.902	0.755
EWB	0.921	0.935	0.643

Note NWW (new ways of working); TI (turnover intention); TNS (technostress), EWB (employee well-being)

Table 9.3 Fornell–Lacker criterion assessment for discriminant validity of constructs related to new ways of working (NWW), technostress (TNS), employee well-being (EWB), and turnover intention (TI)

Constructs	NWW	TI	TNS	EWB
NWW	0.782			
TI	0.074	0.924		
TNS	0.267	0.309	0.869	
EWB	0.764	0.086	0.315	0.802

Note NWW (new ways of working); TI (turnover intention); TNS (technostress), EWB (employee well-being)

Table 9.4 Heterotrait-Monotrait ratio (HTMT) assessment for discriminant validity of constructs related to new ways of working (NWW), technostress (TNS), employee well-being (EWB), and turnover intention (TI)

Constructs	NWW	TI	TNS	EWB
NWW				
TI	0.100			
TNS	0.297	0.340		
EWB	0.830	0.101	0.355	

Note NWW (new ways of working); TI (turnover intention); TNS (technostress), EWB (employee well-being)

9.4.2 Structural Model Analysis and Hypothesis Testing

Associations were ascertained by examining path coefficients and corresponding *p*-values. Following the confirmation of the reliability and validity of the measurement model, and after establishing the overall model fit, the postulated paths and coefficients were assessed through bootstrapping using a sample size of 5000. This approach facilitated subsequent hypothesis testing based on coefficients and *p*-values. Tables 9.5 and 9.6 present the results of the SEM analysis.

Table 9.5 Path coefficient analysis of relationships among new ways of working (NWW), technostress (TNS), employee well-being (EWB), and turnover intention (TI)

Relationship	Path coefficient	t-statistics	*p*-values
NWW → TI	0.000	0.004	0.997
NWW → TNS	0.267	3.791	0.000
TNS → TI	0.313	5.679	0.000
NWW → EWB	0.732	25.194	0.000
EWB → TI	– 0.013	0.144	0.885
TNS → EWB	0.120	3.154	0.000
NWW → TNS → TI	0.084	2.993	0.003
NWW → EWB → TI	– 0.010	0.143	0.886

Note NWW (new ways of working); TI (turnover intention); TNS (technostress), EWB (employee well-being)

Table 9.6 Hypotheses testing results for relationships among new ways of working (NWW), technostress (TNS), employee well-being (EWB), and turnover intention (TI)

Hypotheses	Relationship	Path coefficient	*p*-values	Conclusion
H1	NWW → TI	0.000	0.997	Not supported
H2	NWW → TNS → TI	0.084	0.003	Supported
H3	NWW → EWB → TI	– 0.010	0.886	Not supported
H4	NWW → TNS → EWB → TI	0.000	0.893	Not supported

Note NWW (new ways of working); TI (turnover intention); TNS (technostress), EWB (employee well-being)

The data statistically upheld the positive correlations between NWW and TNS and between NWW and EWB ($\beta = 0.267, p = 0.000$; $\beta = 0.732$, $p = 0.000$), respectively. The dataset did not statistically substantiate the expected negative correlation between NWW and TI ($\beta = 0.00, p = 0.997$), rendering limited evidence for hypothesis H1. We see that there is a positive relationship between TNS and EWB (β = 0.12, p = 0.000). Conversely, the predicted negative correlation between EWB and TI was not statistically significant (β = – 0.013, p = 0.885). Furthermore, TNS demonstrated a statistically significant mediating effect between NWW and TI (β = 0.084, p = 0.003), endorsing hypothesis H2. EWB did not exhibit a significant mediating effect between NWW and TI (β = – 0.010, p = 0.886), indicating that hypothesis H3 lacked support. Finally, TNS and WB did not jointly exhibit a dual mediation effect in the NWW-TI relationship (β = – 0.000, p = 0.893), leading to the non-affirmation of hypothesis H4.

In examining the digital work era within Vietnam's unique socio-cultural context, we found a significant positive correlation between NWW and TNS, aligning with Molino et al. (2020) and Andrulli and Gerards (2023), but diverging from Taser et al. (2022). This suggests that the relationship between NWW and TNS is dynamic and influenced by various factors such as organisational culture, technology implementation, and individual roles within a company. The variability in these findings highlights the importance of contextual factors in shaping how technological advancements are integrated into the workplace and their subsequent effects on employee stress levels.

NWW was also found to significantly enhance EWB, evidenced by a robust statistical correlation. This supports findings from Andrulli and Gerards (2023) and De Vincenzi et al. (2022), contrasting with Lavy (2019). The discrepancy suggests that the impact of NWW on EWB is not uniform and requires specific conditions or support mechanisms for positive outcomes. Tailoring NWW strategies to individual organisational and employee needs is crucial for optimising their impact on EWB. This entails adopting flexible work arrangements and ensuring they are supported by appropriate resources and organisational support structures to foster genuine improvements in well-being.

Contrary to our expectations, our findings indicate an insignificant relationship between NWW and TI, challenging prior studies that reported reduced TI due to flexible and autonomous work practises (Bainbridge & Townsend, 2020; Krajcsák & Kozák, 2022). This highlights Vietnam's unique cultural and organisational dynamics and the need for a nuanced understanding of workplace practices' influence on employee retention within this context. It suggests that while flexible work practices are generally seen as beneficial, their effectiveness in reducing turnover may depend heavily on how they are perceived and implemented within Vietnamese companies' specific cultural and organisational framework.

Surprisingly, we observed a positive correlation between TNS and EWB, contrary to traditional literature (Haar & Brougham, 2022; Suh & Lee, 2017; Wang et al., 2021; Xiao et al., 2023) that posits a negative relationship between these variables. This could indicate that a moderate level of TNS might stimulate personal growth and contribute to well-being or reflect a cultural shift in the perception of TNS in the workplace. The finding challenges conventional wisdom and opens a discussion about the potential benefits of TNS, possibly as a catalyst for employees to develop new skills and resilience, thus enhancing their sense of well-being despite the challenges.

Our analysis also revealed that EWB does not significantly relate to TI in the Vietnamese context. This finding deviates from the common belief that lower well-being is associated with higher TI and may be attributed to unique cultural, economic, or organisational factors in Vietnam. It suggests that factors like job security, financial incentives, or cultural commitments may play a more significant role in an employee's decision to stay or leave an organisation. Additionally, we found that TNS significantly mediates the relationship between NWW and TI, indicating that while NWW does not directly affect TI, TNS influences this relationship. This highlights the importance of TNS in evaluating modern work practices' impacts on employee retention. Finally, the anticipated serial mediation through TNS and WB in the relationship between NWW and TI is not supported.

9.5 Discussion of Implications

9.5.1 Theoretical and Practical Implications

Our research contributes to theoretical understandings by comparing traditional frameworks with digital nuances, emphasising the complexity of securing and protecting valuable resources in the new ways of working (NWW) era. Technostress (TNS) emerges as a critical factor in this landscape, resonating with the COR theory. Our findings expand the theoretical scope of NWW impacts, challenging the notion that NWW straightforwardly leads to improved employee well-being (EWB) or reduced turnover intention (TI). This suggests a need for rethinking theoretical models around NWW and exploring non-linear or simultaneous influences on TI. Practically, our findings imply that merely integrating NWW into organisational practices might not suffice to reduce TI. Mindful, strategic approaches to implementing change and managing TNS are crucial. Moreover, our study highlights the necessity for comprehensive, multi-dimensional retention strategies, recognising that more than well-being initiatives alone may be needed to fully address the complex determinants of employee retention.

In Vietnam's evolving workplace, unique socio-cultural nuances significantly mould employee attitudes and behaviours. This calls for empirical studies deeply rooted in local contexts. The challenges faced with adopting IT in emerging economies underscore the necessity for robust support systems during technological transitions. This study emphasises the necessity of comprehensive technological transformation to prevent societal divides. It aligns with the SDGs by stressing the importance of balancing technical progress with well-being and sustainable economic growth. In summary, our study offers a rich perspective on modern work dynamics, filling knowledge gaps and enhancing theoretical and practical understandings in the context of emerging countries like Vietnam.

9.5.2 Comparison with Findings from Other Asian Countries

This research delineates how NWW manifest across various Asian contexts, highlighting a distinct divergence between developed and developing countries. In developed countries like South Korea, Saudi Arabia, and China the findings are nuanced. South Korea's approach underlines the importance of digital culture, leadership, and knowledge management in bolstering mental well-being, suggesting that comprehensive organisational support can effectively buffer TNS in technologically advanced sectors (Li et al., 2025). Conversely, In Saudi Arabia, research demonstrated how TNS exacerbates work withdrawal, and negatively impacting EWB amid the country's rapid digital transformation (Shamout et al., 2022). The findings from Chinese university teachers revealed that techno-stressors' challenge appraisal is positively related to continuance intention among those teaching from home during the pandemic (Wang et al., 2024).

Conversely, in developing countries such as India and Malaysia, the impact of NWW and TNS exhibits different dynamics. Indian studies on enterprise social network usage increase TNS; TNS increases TI through the mediating role of mental health. Interestingly, perceived organisational support moderated this effect, perceived organisational support reduces employee TI by reinforcing the effect of TNS on employee mental health, suggesting that psychological resources can mitigate TNS's impact on TI (Nayak & Budhwar, 2022). Meanwhile, Malaysian employees report positive experiences with NWW, particularly in task crafting, which appears to benefit short-tenured employees significantly (Ali et al., 2023). Additionally, a study from India highlighted the significant roles of digital technology use in moderating the relationships between mindfulness and psychological well-being. The research underscored that for individuals with low digital technology use, there is a significant optimistic prediction of psychological well-being (Sode et al., 2024).

Lastly, an industry-specific study across Asia's hospitality sector revealed that enhancing digital transformation readiness-including investing in digital skills, fostering positive attitudes toward technology, and ensuring user-friendly systems effectively reduces TNS and supports EWB (Alfehaid et al., 2024). Taken together, these cross-country comparisons suggest that while TNS, EWB, and employee behaviour are common concerns across Asia, the outcomes and mediating factors differ based on cultural norms, organisational support, resilience, and policy frameworks. These insights provide valuable guidance for policymakers and organisations throughout Asia, emphasising the need for tailored strategies that address both technological demands and EWB.

9.6 Future Directions

The findings delve into the dual nature of the new ways of working (NWW), demonstrating its capacity to both elevate technostress (TNS) and enhance employee well-being (EWB) while showing no significant direct effect on turnover intention (TI). This divergence from earlier research suggests that although NWW reshapes employees' psychological experiences, it does not necessarily influence their intent to leave. This insight is crucial for envisioning how organisations across Asia might strategically adapt NWW models as they shape the future of work.

Looking ahead, NWW holds considerable potential if strategically refined to amplify benefits and minimise drawbacks. Organisations and policymakers can reduce TNS and sustain EWB by investing in digital skills training, strengthening technological infrastructure, promoting digital literacy, and reinforcing work-life boundaries. These actions are essential not just for current workforces but also for preparing future-oriented systems that can thrive amid rapid digital transformation in Asia. Managerial support and resilience-building initiatives are also crucial to help employees manage workload pressures and avoid over-connectivity. As flexible and remote work models gain traction across Asia, driven by globalisation, cost efficiency, and rising employee demand for flexibility, more companies are likely to adopt NWW. However, its success depends on critical conditions: robust infrastructure, strong organisational support, clear communication, and active employee participation in policy design. The future success of NWW will rely on how effectively these elements are integrated into evolving work models across diverse Asian contexts. Under these conditions, NWW may evolve into a more effective, employee-centred model that not only promotes well-being but also contributes to lower TI and a more resilient workforce in the digital era.

9.7 Conclusion

Our study offers insights into new ways of working (NWW), technostress (TNS), employee well-being (EWB), and turnover intention (TI) dynamics within Vietnam's unique context. Key findings underscore the complex interplay between these variables, revealing that while NWW can enhance EWB, it simultaneously elevates TNS, influencing TI. Despite the positive impact of NWW on EWB, this enhancement in well-being does not translate into reduced TI, challenging the traditional assumptions about the straightforward benefits of flexible work practises. The findings suggest that the implementation of NWW, although beneficial in fostering a more adaptable and satisfied workforce, must be approached with caution, given its nuanced effects on employee stress levels and potential turnover. This study extends the theoretical discussion on NWW by highlighting its implications for future work in Asia and calling for a more nuanced understanding of how emerging work models are implemented and perceived across diverse cultural contexts. Moreover, the significant relationships between TNS, EWB and TI emphasise the need for organisational strategies that promote flexible working conditions and effectively address stressors. By integrating these insights into human resource policies, organisations can better manage the dichotomy between enhancing employee autonomy and mitigating stress-related consequences. In conclusion, this research contributes to a deeper understanding of the adaptive challenges and opportunities presented by NWW in an emerging market context like Vietnam. It advocates for strategic, future-oriented, and culturally informed approaches to adopting NWW practices, aiming to maximise their long-term benefits while minimising potential negative consequences on retention and workforce sustainability in the digital age. Future research should continue to explore these dynamics across diverse cultural landscapes to refine our global comprehension of the evolving nature of work.

Highlights

- New ways of working are shaping the future of work in Asia, simultaneously enhancing employee well-being and increasing technostress, but do not directly reduce turnover intention.
- Technostress acts as a key mediator in understanding how digital work models will influence future workforce dynamics across Asian contexts.
- Turnover intentions in Asian countries, including but not limited to Vietnam, are influenced by diverse cultural and economic factors, underscoring the need for region-specific workforce strategies.
- Flexible work strategies must evolve into future-ready models that respond to the unique socio-cultural dynamics of Asia, supporting both employee resilience and digital transformation.

References

Akbari, M., & Hopkins, J. L. (2022). Digital technologies as enablers of supply chain sustainability in an emerging economy. *Operations Management Research, 15*(3), 689–710. https://doi.org/10.1007/s12063-021-00226-8

Alfehaid, M. M., Hasan, T. H., Alfisal, A. A., Saleh, M. I., Bumbak, S. V., & Helal, M. Y. (2024). Post-pandemic digital transformation: Mitigating technostress and increasing well-being in the hospitality industry via the person-technology fit model. *Geo Journal of Tourism and Geosites, 53*(2), 388–399. https://doi.org/10.30892/gtg.53202-1214

Alfes, K., Avgoustaki, A., Beauregard, T. A., Cañibano, A., & Muratbekova-Touron, M. (2022). New ways of working and the implications for employees: A systematic framework and suggestions for future research. *The International Journal of Human Resource Management, 33*(22), 4361–4385. https://doi.org/10.1080/09585192.2022.2149151

Ali, S. N. M., Cioca, L.-I., Kayati, R. S., Saputra, J., Adam, M., Plesa, R., & Ibrahim, R. Z. A. R. (2023). A study of psychometric instruments and constructs of work-related stress among seafarers: A qualitative approach. *International Journal of Environmental Research and Public Health, 20*(4), 2866. https://doi.org/10.3390/ijerph20042866

Alpar, P. (2020). Turnover intentions of employees of information technology outsourcing suppliers in Vietnam. *International Journal of Human Resources Development and Management, 20*(1), 43–60. https://doi.org/10.1504/IJHRDM.2020.105109

Andrulli, R., & Gerards, R. (2023). How new ways of working during COVID-19 affect employee well-being via technostress, need for recovery, and work engagement. *Computers in Human Behavior, 139*, Article 107560. https://doi.org/10.1016/j.chb.2022.107560

Angelici, M., & Profeta, P. (2023). Smart working: work flexibility without constraints. *Management Science*. https://doi.org/10.1287/mnsc.2023.4767

Bainbridge, H. T., & Townsend, K. (2020). The effects of offering flexible work practices to employees with unpaid caregiving responsibilities for elderly or disabled family members. *Human Resource Management, 59*(5), 483–495. https://doi.org/10.1002/hrm.22007

Bakker, A. B., Demerouti, E., & Sanz-Vergel, A. (2023). Job demands–resources theory: Ten years later. *Annual Review of Organizational Psychology and Organizational Behavior, 10*(1), 25–53. https://doi.org/10.1146/annurev-orgpsych-120920-053933

Bellmann, L., & Hübler, O. (2021). Working from home, job satisfaction and work–life balance—Robust or heterogeneous links? *International Journal of Manpower, 42*(3), 424–441. https://doi.org/10.1108/IJM-10-2019-0458

Brislin, R. W. (1976). Comparative research methodology: Cross-cultural studies. *International Journal of Psychology, 11*(3), 215–229. https://doi.org/10.1080/00207597608247359

Burns, R. A., Butterworth, P., & Anstey, K. J. (2016). An examination of the long-term impact of job strain on mental health and wellbeing over a 12-year period. *Social Psychiatry and Psychiatric Epidemiology, 51*(5), 725–733. https://doi.org/10.1007/s00127-016-1192-9

Caillier, J. G. (2018). Do flexible work schedules reduce turnover in U.S. federal agencies? *The Social Science Journal*, *55*(2), 108–115. https://doi.org/10.1016/j.soscij.2017.09.005

Campbell, M., & Gavett, G. (2021). What covid-19 has done to our well-being, in 12 charts. *Harvard Business Review, 10.*

Cao, M., Zhao, Y., & Zhao, S. (2023). How CEOs' inclusive leadership fuels employees' well-being: A three-level model. *The International Journal of Human Resource Management, 34*(11), 2305–2330. https://doi.org/10.1080/09585192.2022.2077126

Chillakuri, B., & Vanka, S. (2021). Examining the effects of workplace well-being and high-performance work systems on health harm: A Sustainable HRM perspective. *Society and Business Review, 16*(1), 71–93. https://doi.org/10.1108/SBR-03-2020-0033

De Vincenzi, C., Pansini, M., Ferrara, B., Buonomo, I., & Benevene, P. (2022). Consequences of COVID-19 on employees in remote working: Challenges, risks and opportunities an evidence-based literature review. *International Journal of Environmental Research and Public Health, 19*(18), 11672. https://www.mdpi.com/1660-4601/19/18/11672

Demerouti, E., Bakker, A. B., Nachreiner, F., & Schaufeli, W. B. (2001). The job demands-resources model of burnout. *Journal of Applied Psychology, 86*(3), 499. https://doi.org/10.1037/0021-9010.86.3.499

Dysvik, A., & Kuvaas, B. (2013). Perceived job autonomy and turnover intention: The moderating role of perceived supervisor support. *European Journal of Work and Organizational Psychology, 22*(5), 563–573. https://doi.org/10.1080/1359432X.2012.667215

Fornell, C., & Larcker, D. F. (1981). Structural equation models with unobservable variables and measurement error: Algebra and statistics. *Journal of Marketing Research, 18*(3), 382–388. https://doi.org/10.1177/002224378101800313

Gajendran, R. S., & Harrison, D. A. (2007). The good, the bad, and the unknown about telecommuting: Meta-analysis of psychological mediators and individual consequences. *Journal of Applied Psychology, 92*(6), 1524. https://doi.org/10.1037/0021-9010.92.6.1524

Gerards, R., de Grip, A., & Baudewijns, C. (2018). Do new ways of working increase work engagement? *Personnel Review, 47*(2), 517–534. https://doi.org/10.1108/PR-02-2017-0050

Gerards, R., de Grip, A., & Weustink, A. (2021). Do new ways of working increase informal learning at work? *Personnel Review, 50*(4), 1200–1215. https://doi.org/10.1108/PR-10-2019-0549

Gillespie, N. A., Walsh, M., Winefield, A. H., Dua, J., & Stough, C. (2001). Occupational stress in universities: Staff perceptions of the causes, consequences and moderators of stress. *Work & Stress, 15*(1), 53–72. https://doi.org/10.1080/02678370117944

Haar, J., & Brougham, D. (2022). Work antecedents and consequences of work-life balance: A two sample study within New Zealand. *The International Journal of Human Resource Management, 33*(4), 784–807. https://doi.org/10.1080/09585192.2020.1751238

Hair, J., Hult, G. T. M., Ringle, C., & Sarstedt, M. (2021). *A primer on partial least squares structural equation modeling (PLS-SEM).* Sage Publications.

Hair, J. F., Anderson, R. E., Babin, B. J., & Black, W. C. (2010). *Multivariate data analysis: A global perspective* (Vol. 7). Pearson.

Hair, J. F., Risher, J. J., Sarstedt, M., & Ringle, C. M. (2019). When to use and how to report the results of PLS-SEM. *European Business Review, 31*(1), 2–24. https://doi.org/10.1108/EBR-11-2018-0203

Han, J. W. (2023). A contextual study of employee turnover intention in Vietnam F&B service sector: An integrative perspective. *Journal of Advances in Management Research, 20*(1), 35–54. https://doi.org/10.1108/JAMR-04-2022-0079

Harris, K. J., Harris, R. B., Valle, M., Carlson, J., Carlson, D. S., Zivnuska, S., & Wiley, B. (2022). Technostress and the entitled employee: Impacts on work and family. *Information Technology & People, 35*(3), 1073–1095. https://doi.org/10.1108/ITP-07-2019-0348

Henseler, J., Ringle, C. M., & Sarstedt, M. (2015). A new criterion for assessing discriminant validity in variance-based structural equation modeling. *Journal of the Academy of Marketing Science, 43*(1), 115–135. https://doi.org/10.1007/s11747-014-0403-8

Hobfoll, S. E. (1989). Conservation of resources: A new attempt at conceptualizing stress. *American Psychologist, 44*(3), 513–524. https://doi.org/10.1037/0003-066X.44.3.513

Hobfoll, S. E. (2001). The influence of culture, community, and the nested-self in the stress process: Advancing conservation of resources theory. *Applied Psychology, 50*(3), 337–421. https://doi.org/10.1111/1464-0597.00062

Hofstede, G. (1993). Cultural constraints in management theories. *Academy of Management Perspectives, 7*(1), 81–94. https://doi.org/10.5465/ame.1993.9409142061

Hwang, I., & Cha, O. (2018). Examining technostress creators and role stress as potential threats to employees' information security compliance. *Computers in Human Behavior, 81*, 282–293. https://doi.org/10.1016/j.chb.2017.12.022

Jaiswal, A., Sengupta, S., Panda, M., Hati, L., Prikshat, V., Patel, P., & Mohyuddin, S. (2022). Teleworking: Role of psychological well-being and technostress in the relationship between trust in management and employee performance. *International Journal of Manpower*. https://doi.org/10.1108/IJM-04-2022-0149

Jamal, M. T., Anwar, I., Khan, N. A., & Ahmad, G. (2023). How do teleworkers escape burnout? A moderated-mediation model of the job demands and turnover intention. *International Journal of Manpower*. https://doi.org/10.1108/IJM-12-2022-0628

Jang, J., & Kandampully, J. (2018). Reducing employee turnover intention through servant leadership in the restaurant context: A mediation study of affective organizational commitment. *International Journal of Hospitality & Tourism Administration, 19*(2), 125–141. https://doi.org/10.1080/15256480.2017.1305310

Jensen, M. T. (2021). Pupil-teacher ratio, disciplinary problems, classroom emotional climate, and turnover intention: Evidence from a randomized control trial. *Teaching and Teacher Education, 105*, Article 103415. https://doi.org/10.1016/j.tate.2021.103415

Ketkaew, C., Manglakakeeree, O., & Naruetharadhol, P. (2020). The interrelationships of work-related factors, person-environment fit, and employee turnover intention. *Cogent Business & Management, 7*(1), 1823580. https://doi.org/10.1080/23311975.2020.1823580

Krajcsák, Z., & Kozák, A. (2022). The moderating role of remote work in the relationship between organizational culture and OCB: Case studies from the financial sector. *Journal of Advances in Management Research, 19*(2), 300–315. https://doi.org/10.1108/JAMR-07-2021-0247

Lavy, S. (2019). Daily dynamics of teachers' organizational citizenship behavior: Social and emotional antecedents and outcomes. *Frontiers in Psychology, 10*, 2863. https://doi.org/10.3389/fpsyg.2019.02863

Le, H., Lee, J., Nielsen, I., & Nguyen, T. L. A. (2022). Turnover intentions: the roles of job satisfaction and family support. *Personnel Review*. https://doi.org/10.1108/PR-08-2021-0582

Li, X., Seah, R. Y. T., & Yuen, K. F. (2025). Mental wellbeing in digital workplaces: The role of digital resources, technostress, and burnout. *Technology in Society, 81*, Article 102844. https://doi.org/10.1016/j.techsoc.2025.102844

Mahomed, F., Oba, P., & Sony, M. (2023). Exploring employee well-being during the COVID-19 remote work: Evidence from South Africa. *European Journal of Training and Development, 47*(10), 91–111. https://doi.org/10.1108/EJTD-06-2022-0061

Maier, C., Laumer, S., & Weitzel, T. (2022). A dark side of telework: A social comparison-based study from the perspective of office workers. *Business & Information Systems Engineering, 64*(6), 793–811. https://doi.org/10.1007/s12599-022-00758-8

Metselaar, S. A., den Dulk, L., & Vermeeren, B. (2023). Teleworking at different locations outside the office: Consequences for perceived performance and the mediating role of autonomy and work-life balance satisfaction. *Review of Public Personnel Administration, 43*(3), 456–478. https://doi.org/10.1177/0734371X221087421

Molino, M., Ingusci, E., Signore, F., Manuti, A., Giancaspro, M. L., Russo, V., Zito, M., & Cortese, C. G. (2020). Wellbeing costs of technology use during Covid-19 remote working: An investigation using the Italian translation of the technostress creators scale. *Sustainability, 12*(15), 5911. https://doi.org/10.3390/su12155911

Narahari, C. L., & Koneru, D. (2017). A study on the role of occupational stress in organisations. *International Journal of Engineering Technology, Management and Applied Sciences, 5*(1), 53–59.

Nastjuk, I., Trang, S., Grummeck-Braamt, J.-V., Adam, M. T., & Tarafdar, M. (2023). Integrating and synthesising technostress research: A meta-analysis on technostress creators, outcomes, and IS usage contexts. *European Journal of Information Systems*, 1–22. https://doi.org/10.1080/0960085X.2022.2154712

Nayak, S., & Budhwar, P. (2022). Negative effects of enterprise social networks (ESNs) and technostress: Empirical evidence from R&D centres operating in India. *Aslib Journal of Information Management, 74*(5), 956–988. https://doi.org/10.1108/AJIM-08-2021-0228

Ollo-López, A., Goñi-Legaz, S., & Erro-Garcés, A. (2021). Home-based telework: Usefulness and facilitators. *International Journal of Manpower, 42*(4), 644–660. https://doi.org/10.1108/IJM-02-2020-0062

Puhakka, I. J. A., Nokelainen, P., & Pylväs, L. (2021). Learning or leaving? Individual and environmental factors related to job satisfaction and turnover intention. *Vocations and Learning, 14*(3), 481–510. https://doi.org/10.1007/s12186-021-09275-3

Rubenstein, A. L., Eberly, M. B., Lee, T. W., & Mitchell, T. R. (2018). Surveying the forest: A meta-analysis, moderator investigation, and future-oriented discussion of the antecedents of voluntary employee turnover. *Personnel Psychology, 71*(1), 23–65. https://doi.org/10.1111/peps.12226

Shamout, M. D., Elayan, M. B. H., Hamouche, S., Rawashdeh, A. M., & Elrehail, H. (2022). Impact of technostress on withdrawal behavior and workplace flourishing: Do contextual variables matter? *Information Resources Management Journal (IRMJ), 35*(1), 1–17. https://doi.org/10.4018/IRMJ.312212

Shanmugam, M. M., & Agarwal, B. (2019). Support perceptions, flexible work options and career outcomes. *Gender in Management: An International Journal, 34*(4), 254–286. https://doi.org/10.1108/GM-12-2018-0157

Sode, R., Chenji, K., & Vijayaraghavan, R. (2024). Exploring workplace spirituality, mindfulness, digital technology, and psychological well-being: A complex interplay in organizational contexts. *Acta Psychologica, 251*, Article 104601. https://doi.org/10.1016/j.actpsy.2024.104601

Suh, A., & Lee, J. (2017). Understanding teleworkers' technostress and its influence on job satisfaction. *Internet Research, 27*(1), 140–159. https://doi.org/10.1108/IntR-06-2015-0181

Tarafdar, M., Tu, Q., Ragu-Nathan, B. S., & Ragu-Nathan, T. S. (2007). The impact of technostress on role stress and productivity. *Journal of Management Information Systems, 24*(1), 301–328. https://doi.org/10.2753/MIS0742-1222240109

Tarafdar, M., Tu, Q., & Ragu-Nathan, T. S. (2010). Impact of technostress on end-user satisfaction and performance. *Journal of Management Information Systems, 27*(3), 303–334. https://doi.org/10.2753/MIS0742-1222270311

Taser, D., Aydin, E., Torgaloz, A. O., & Rofcanin, Y. (2022). An examination of remote e-working and flow experience: The role of technostress and loneliness. *Computers in Human Behavior, 127*, Article 107020. https://doi.org/10.1016/j.chb.2021.107020

Wang, Q., Zhao, G., & Yao, N. (2024). Understanding the impact of technostress on university teachers' online teaching during the COVID-19 pandemic with the Transactional Theory of Stress (TTS). *The Asia-Pacific Education Researcher, 33*(1), 187–198. https://doi.org/10.1007/s40299-023-00718-0

Wang, Z., Xu, H., & Song, M. (2021). Exploring how and when ethical conflict impairs employee organizational commitment: A stress perspective investigation. *Business Ethics, the Environment & Responsibility, 30*(2), 172–187. https://doi.org/10.1111/beer.12324

Xiao, S.-P., Lu, Y., Yan, Y., Zhou, Z. E., Cao, Z.-X., & Zhao, K.-C. (2023). Effects of two-dimensional cyber incivility on employee well-being from a self-determination perspective. *Frontiers in Psychology, 14*, 1137587. https://doi.org/10.3389/fpsyg.2023.1137587

Yadav, L. K. (2016). Employee engagement among academicians: Interaction effect of perceived organizational support and individualism. *Vilakshan: The XIMB Journal of Management, 13*(1).

Yasin, R., Jan, G., Huseynova, A., & Atif, M. (2023). Inclusive leadership and turnover intention: The role of follower–leader goal congruence and organizational commitment. *Management Decision, 61*(3), 589–609. https://doi.org/10.1108/MD-07-2021-0925

Zhu, Y., Warner, M., & Rowley, C. (2013). Human resource management with 'Asian' characteristics: A hybrid people-management system in East Asia. In *Globalizing international human resource management* (pp. 29–52). Routledge.

Phuong Ngoc-Duy Nguyen is Associate Professor in Economics and researcher at the School of Economics, Finance & Accounting, International University, Vietnam National University Ho Chi Minh City. He earned his Ph.D. in Development Administration from the National Institute of Development Administration in 2013. His major research and teaching interests focus on public policy, digital entrepreneurship, sustainable development, corporate strategy, circular economy, customer and employee behaviour on OCB, commitment, and retention, and he has published over 100 research articles.

Cong Hiep Duong is a Lecturer at the Thai Nguyen University of Economics and Business Administration, Thai Nguyen, Vietnam. He earned his Ph.D. in Business and Operations Management from the Chang Jung Christian University in 2024. His research focuses on organisational behaviour and human resource management. With his dedication, expertise, and passion, he is poised to make contributions that will shape the future of the academic and business communities.

Thi Trang Giang earned her Ph.D. in Business and Operations Management from Chang Jung Christian University in Taiwan. She is currently a lecturer at Thai Nguyen University of Economics and Business Administration in Vietnam. Her research interests include organisational behaviour, human resource management, and business ethics. She has contributed to several academic publications, including articles in the Journal of Organisational Analysis, Sage Open, and the International Journal of Educational Management.

Huan Hong Nguyen is a dedicated young researcher with a strong blend of theoretical knowledge and practical expertise. With an MBA and an MSc in Computer Science, Huan has also earned the esteemed Project Management Professional (PMP®) certification, demonstrating his capability in project management and leadership. His involvement in various publications, his research and data analysis skills, and his teaching experience reflect his commitment to contributing meaningfully to technology management and human development.

Index

C. Y. Fung et al. (eds.), *Future of Work in Asia*,
https://doi.org/10.1007/978-3-031-95804-5

MIX
Papier aus verantwortungsvollen Quellen
Paper from responsible sources
FSC® C105338

If you have any concerns about our products,
you can contact us on
ProductSafety@springernature.com

In case Publisher is established outside the EU,
the EU authorized representative is:
Springer Nature Customer Service Center GmbH
Europaplatz 3, 69115 Heidelberg, Germany

Printed by Libri Plureos GmbH
in Hamburg, Germany